\# 39-90 BKBud June 90

NORTH CAROLINA
STATE BOARD OF COMMUNITY COLLEGES
LIBRARIES
SOUTHEASTERN COMMUNITY COLLEGE

D1708947

For Reference

Not to be taken from this room

SOUTHEASTERN COMMUNITY
COLLEGE LIBRARY
WHITEVILLE, NC 28472

Poisonous Plants
of Eastern North America

POISONOUS PLANTS

Of Eastern
North
America

By
Randy G. Westbrooks
James W. Preacher

UNIVERSITY OF SOUTH CAROLINA PRESS

Copyright © University of South Carolina 1986

First Edition

Published in Columbia, South Carolina, by the
University of South Carolina Press

Printed in Hong Kong by
Everbest Printing Company, Ltd., for
Four Colour Imports, Ltd. – Louisville

Library of Congress Cataloging-in-Publication Data

Westbrooks, Randy G., 1953–
  Poisonous plants of eastern North America.

  Bibliography: p.
  Includes index.
  1. Poisonous plants—United States—Identification.
  2. Poisonous plants—Canada—Identification
  3. Poisonous plants—Toxicology. I. Preacher, James W.,
1949– . II. Title.
QK100.U6W47 1986   581.6'9'097   86–1669
ISBN 0-87249-442-X

PHOTO CREDITS

    Sincere appreciation is extended to the following people and organizations who graciously provided use of their plant slides for use in this publication:

Dr. W. T. Batson, Professor Emeritus, University of South Carolina, Columbia: *Ornithogalum umbellatum*, p. 29.

Dr. Ritchie Bell, University of North Carolina, Chapel Hill: *Momordica charantia*, p. 165; *Pastinaca sativa*, p. 132; *Symplocarpus foetidus*, p. 19.

Ms. Nancy Hammer and Fairchild Tropical Garden, Miami, Florida: *Solandra maxima*, p. 158; *Hura crepitans*, p. 102.

Pamela Harper and Harper Horticultural Slide Library, Seaford, Virginia: *Atropa belladonna*, p. 148; *Colchium autumnale*, p. 23; *Lathyrus spp.*, p. 82.

Dr. William S. Justice, deceased, Asheville, North Carolina: *Agrostemma githago*, p. 52; *Caltha palustris*, p. 57; *Hydrastis canadensis*, p. 61; *Veratrum viride*, p. 30.

Dr. R. J. Knight, Jr. and Subtropical Horticulture Research Station, USDA ARS, Miami, Florida: *Cestrum nocturnum*, p. 151; *Gloriosa superba*, p. 26.

Dr. Erik A. Neumann, U.S. National Arboretum, Washington, D.C.: *Laburnum anagyroides*, p. 81.

James B. Watson and Fairchild Tropical Garden, Miami, Florida: *Metopium toxiferum*, p. 109.

To our wives,
Rebecca Marks Westbrooks
and
Carol AnLaurie Preacher

and to our sons,
Matthew and *John Westbrooks*
and
Jeffery and *Troy Preacher*

Contents

(*text species grouped by families*)

- xiii   PREFACE
- xv   FOREWORD
- xvii   INTRODUCTION
- xviii   ACKNOWLEDGMENTS
- xix   PLANT TOXINS
- 1   PICTORIAL GLOSSARY

CYCADACEAE
- 3   *Cycas circinalis* (false sago palm)
- 5   *Zamia spp.* (coontie)

GINKGOACEAE
- 6   *Ginkgo biloba*

CUPRESSACEAE
- 7   *Juniperus spp.* (red cedar)

TAXACEAE
- 8   *Taxus spp.* (yew)

PODOCARPACEAE
- 9   *Podocarpus macrophylla* (Japanese yew)
- 10   *Caryota mitis* (fishtail palm)

ARACEAE
- 11   *Arisaema triphyllum* (jack-in-the-pulpit)
- 12   *Caladium spp.*
- 13   *Dieffenbachia spp.* (dumbcane)
- 14   *Epipremum aureum* (hunter's robe)
- 15   *Monstera deliciosa* (split-leaf philodendron)
- 16   *Orontium aquaticum* (golden club)
- 17   *Philodendron spp.*
- 18   *Pistia stratiotes* (water lettuce)
- 19   *Symplocarpus foetidus* (skunk cabbage)

COMMELINACEAE
- 20   *Rhoeo spathacea* (oyster plant)
- 21   *Tradescantia pallida* (purple queen)

LILIACEAE
- 22   *Asparagus officinalis*
- 23   *Colchium autumnale* (autumn crocus)
- 25   *Convallaria majalis* (lily-of-the-valley)
- 26   *Gloriosa superba* (glory lily)
- 28   *Hyacinthus orientalis* (hyacinth)
- 29   *Ornithogalum umbellatum* (star-of-Bethlehem)
- 30   *Veratrum viride* (false hellebore)
- 32   *Zigadenus densus* (black snakeroot)

Contents

AMARYLLIDACEAE
- 33 *Agave spp.* (century plant)
- 34 *Amaryllis spp.*
- 35 *Crinum spp.* (milk & wine lily)
- 36 *Narcissus spp.* (jonquil, daffodil)

IRIDACEAE
- 37 *Iris spp.*

ORCHIDACEAE
- 38 *Cypripedium spp.* (lady's slipper orchid)

MORACEAE
- 39 *Ficus carica* (fig)
- 40 *Morus rubra* (mulberry)

URTICACEAE
- 41 *Urtica dioica* (stinging nettle)

CANNABACEAE
- 42 *Cannabis sativa* (marijuana)
- 43 *Humulus lupulus* (hops)

LORANTHACEAE
- 44 *Phoradendron serotinum* (mistletoe)

POLYGONACEAE
- 45 *Fagopyrum sagittatum* (buckwheat)
- 46 *Rheum rhaponticum* (rhubarb)
- 47 *Rumex spp.* (sorrel)

CHENOPODIACEAE
- 48 *Chenopodium ambrosioides* (Mexican tea)

NYCTAGINACEAE
- 49 *Mirabilis jalapa* (four o'clock)

PHYTOLACCACEAE
- 50 *Phytolacca americana* (pokeberry)

CARYOPHYLLACEAE
- 52 *Agrostemma githago* (corn cockle)
- 53 *Saponaria officinalis* (bouncing bet)

RANUNCULACEAE
- 54 *Aconitum spp.* (monkshood)
- 55 *Actea spp.* (baneberry)
- 56 *Anemone spp.* (wind flower)
- 57 *Caltha palustris* (marsh marigold, cowslip)
- 58 *Clematis spp.* (virgin's bower)
- 59 *Delphinium spp.* (larkspur)
- 60 *Helleborus niger* (christmas rose)
- 61 *Hydrastis canadensis* (golden seal)
- 62 *Ranunculus spp.* (buttercup)

ANNONACEAE
- 63 *Asimina triloba* (pawpaw)

BERBERIDACEAE
- 64 *Caulophyllum thalictroides* (blue cohosh)
- 65 *Podophyllum peltatum* (mayapple)

Contents

MENISPERMACEAE
67 *Menispermum canadense* (moonseed)

PAPAVERACEAE
68 *Argemone mexicana* (Mexican prickle poppy)
69 *Chelidonium majus* (celandine)
70 *Papaver somniferum* (poppy)
71 *Sanguinaria canadensis* (bloodroot)

FUMARIACEAE
72 *Dicentra spp.* (dutchman's breeches)

SAXIFRAGACEAE
73 *Hydrangea spp.*

ROSACEAE
74 *Malus sylvestris* (apple)
75 *Prunus spp.* (cherry)

LEGUMINOSEAE
77 *Abrus precatorius* (rosary pea)
78 *Cassia occidentalis* (coffee senna)
79 *Crotolaria spp.* (rattlebox)
80 *Gymnocladus dioica* (Kentucky coffee tree)
81 *Laburnum anagyroides* (golden chain)
82 *Lathyrus spp.* (chick pea)
84 *Lupinus spp.* (lupine)
85 *Robinia pseudoacacia* (black locust)
86 *Sesbania punicea* (purple sesbania)
87 *Tephrosia virginiana* (goat's rue)
88 *Vicia faba* (fava bean)
90 *Wisteria spp.*

LINACEAE
91 *Linum usitatissimum* (flax)

RUTACEAE
92 *Citrus aurantifolia* (lime)
93 *Poncirus trifoliata* (trifoliate orange)
94 *Ruta graveolens* (rue)

MELIACEAE
95 *Melia azedarach* (chinaberry)

EUPHORBIACEAE
96 *Aleurites fordii* (tung oil tree)
97 *Cnidoscolus stimulosus* (spurge nettle)
98 *Croton capitatus* (woolly croton)
99 *Euphorbia spp.* (spurge)
101 *Hippomane mancinella* (manchineel)
102 *Hura crepitans* (sandbox tree)
103 *Jatropha curcas* (physic nut)
104 *Ricinus communis* (castor bean)
106 *Sapium sebiferum* (popcorn tree)

BUXACEAE
107 *Buxus sempervirens* (boxwood)

Contents

ANACARDIACEAE
108 *Mangifera indica* (mango)
109 *Metopium toxiferum* (poisonwood)
110 *Schinus terebinthifolius* (Brazilian pepper)
111 *Toxicodendron spp.* (poison ivy, oak, sumac)
AQUIFOLIACEAE
113 *Ilex spp.* (holly)
CELASTRACEAE
114 *Euonymous atropurpureus* (strawberry bush)
HIPPOCASTANACEAE
115 *Aesculus spp.* (buckeye)
SAPINADCEAE
117 *Blighia sapida* (akee)
RHAMNACEAE
118 *Rhamnus spp.* (buckthorn)
VITACEAE
120 *Parthenocissus quinquefolia* (Virginia creeper)
HYPERICACEAE
121 *Calophyllum inophyllum* (mastwood)
THYMELACEAE
122 *Dirca palustris* (wicopy, leatherwood)
MYRTACEAE
123 *Eucalyptus spp.*
124 *Melaleuca quinquenervia* (bottlebrush tree)
ARALIACEAE
125 *Aralia spinosa* (hercules club)
126 *Hedera helix* (English ivy)
127 *Polyscias spp.* (aralia)
APIACEAE
128 *Cicuta maculata* (water hemlock)
130 *Conium maculatum* (poison hemlock)
131 *Daucus carota* (Queen Anne's lace)
132 *Pastinaca sativa* (wild parsnip)
ERICACEAE
133 *Kalmia latifolia* (mountain laurel)
134 *Rhododendron spp.* (rhododendron, azalea)
PRIMULACEAE
135 *Primula obconica* (primrose)
OLEACEAE
136 *Ligustrum vulgare* (privet)
LOGANIACEAE
137 *Gelsemium sempervirens* (Carolina jessamine)
139 *Spigelia marilandica* (Indian pink)
APOCYNACEAE
140 *Allamanda Cathartica* (yellow allamanda)
141 *Catharanthus roseus* (periwinkle)
142 *Nerium oleander* (oleander)

ASCLEPIADACEAE
143 *Asclepias spp.* (milkweed)
144 *Cryptostegia grandiflora* (palay rubbervine)

CONVULVULACEAE
145 *Ipomoea tricolor* (morning glory)

VERBENACEAE
146 *Lantana camara*

SOLANACEAE
148 *Atropa belladonna* (belladonna)
150 *Capsicum frutescens* (chili pepper)
151 *Cestrum spp.*
152 *Datura stramonium* (jimsonweed)
154 *Hyoscyamus niger* (henbane)
155 *Lycopersicon esculentum* (tomato)
156 *Nicotiana tabacum* L. (tobacco)
158 *Solandra spp.* (trumpet flower)
159 *Solanum spp.* (nightshade)

SCROPHULARIACEAE
161 *Digitalis purpurea* (foxglove)

BIGNONIACEAE
162 *Campsis radicans* (trumpet creeper)

CAPRIFOLIACEAE
163 *Lonicera japonica* (Japanese honeysuckle)
164 *Sambucus spp.* (elderberry)

CUCURBITACEAE
165 *Momordica charantia* (balsam pear)

LOBELIACEAE
166 *Lobelia spp.* (cardinal flower)

ASTERACEAE
167 *Achillea millefolium* (milfoil, yarrow)
168 *Ageratina altissima* (white snakeroot)
169 *Helenium spp.* (sneezeweed)
170 *Senecio spp.*
172 *Tanacetum vulgare* (tansy)

173 MISCELLANEOUS
183 READY REFERENCE LIST
208 GLOSSARY
213 LITERATURE CITED
221 GENERAL REFERENCES
222 INDEX OF COMMON NAMES

Preface

Unlike the situation earlier when people lived in frontier or agrarian societies and were acquainted, in a practical way, with their natural surroundings, today's urban dwellers, and particularly their children, have little or no contact with or knowledge of such. Whether a plant is useful, harmful or useless is likely of minimum concern with the result that an increasing incidence of difficulty with harmful plants may be expected.

Harmful plants are all around us—in our homes, yards, gardens, floral arrangements, fields, margins and woods. Prevention of harm is better than cure and acquaintance with plants is the first step toward prevention. Parents, teachers, camp counselors, scout leaders and health officials share in the responsibility to protect one another, and especially the young.

Quite a number of poisonous plants just happen to be desirable as ornamentals and are frequently planted. In these cases, the planter's safeguard seems to be familiarity with the plant and its properties and accounts for the responsibility that he consciously assumes in the planting. Later, if or when the ownership of the plants is transferred along with the title to the real estate, as is commonly the case, a warning of any such planting would seem to be most in order.

Contact with or ingestion of plants or plant parts known to be or suspected of being harmful is a matter deserving immediate attention and where a course of action is in doubt contact should be made with the regional Poison Control Center.

This book presents a wide ranging spectrum of botanical and toxicological information. Color photographs of living plants with brief lay language descriptions make for more positive identification. This is followed by toxicity, symptoms and notes with case histories. Common and updated scientific names are given as well as a pictorial glossary of plant parts, a word glossary and complete literature citations. All together, it is a well balanced blend of technical and popular information that will be useful to a variety of readers.

—Wade T. Batson
Professor Emeritus
Department of Biology
University of South Carolina

Foreword

The risk of exposure to plant poisons is not limited to those living in developing countries or farming communities. People in urban and suburban areas are also surrounded by plants. Indoor plants decorate our homes, schools and offices. We landscape around buildings and plan gardens and parks. Wild plants thrive in vacant lots, along fence rows and on undeveloped land. The risks of exposure to plants is evidenced by the fact that this category constitutes the third largest volume of calls to poison control centers in the United States.

Those exposed to poisonous plants risk developing toxic symptoms. Plant toxicity is surrounded by confusion, misunderstanding and gaps in the knowledge base. A large part of the problem is caused by the variability in response to an exposure. Some of the factors responsible for reported variations in toxicity are related to the plant. These include the species of plant, the plant part (leaf, root, fruit, etc.), the amount of plant material involved, the stage of development of the plant and the soil type and growing conditions. Human variables such as age, weight, route of exposure, prior health conditions and individual susceptibility also contribute to differences in responses. In order to estimate the potential for toxic effects and to recommend treatment, poison centers need to know both the name of the plant and as many of the variable factors as possible.

This book should be of value in reducing exposures to poisonous plants and can aid in the assessment of toxicity by identifying poisonous plants. Over 60 percent of reported poison exposures involve young children. Parents, grandparents, day care centers, schools, churches and others who care for children can use the book to identify dangerous plants and make sure children cannot come in contact with them. Sportsmen, scouts, naturalists and others who spend time outdoors can discover plants they should avoid. Health care professionals in doctors' offices, clinics, emergency medical facilities and poison control centers will find the color plates useful in identifying plant material brought in following exposures. The book can be useful even when a plant cannot be found within. The plant description section can be used to describe in correct botanical terms a plant in question. Most poison centers have botanist consultants who can help identify plants from accurate descriptions.

Poisonous Plants of Eastern North America satisfies a real need and will be a valuable addition to many collections.

—*Brooks C. Metts*, Pharm.D.
Palmetto Poison Center
Columbia, South Carolina

Introduction

The nature of edible, herbal and toxic plants was discovered through trial and error and played an important role in man's early survival. Hippocrates, Dioscorides, a Greek Army surgeon, and Galen, a second-century Roman physician, were early scholars on the subject.

Even in urban areas today, the danger of toxic plants still exists because of the increase in such recreational activities as camping and hiking and the popularity of houseplants and gardening. Some common toxic ornamentals include oleander, daphne, philodendron and dumbcane. Tomato vines and apple seeds are also poisonous.

While most people do not chew on parts of unknown plants, children will. Without close supervision, this innocent behavior and their low body weight can lead to serious or even fatal results. It is a good idea to make a list of poisonous plants around the house and yard. Knowing this and the poisonous principles involved can be very helpful in an emergency.

Extreme caution must be observed when gathering wild plants for the table. Many cases of mistaken identity have resulted in poisoning and even death. Herbal species pose a risk as well. Some people believe that if a single dose of herbal medicine is effective, then a double dose will work twice as well. This is dangerous, especially with plants where a critical dosage can mean the difference between medicine and poison. A good example is digitalis, a cardiac drug made from purple foxglove (*Digitalis purpurea* L.). People should consult a physician before taking such preparations.

This book is a literature review intended to advise parents, physicians, poison control centers and others of vascular plants, which are poisonous to humans, that are native to, introduced, or cultivated in eastern North America. Besides those presented here, many other plants are known to be toxic to animals. Many of them are probably toxic to man as well, but no cases have been noted.

Susceptibility to poisonous plants varies considerably among the general populace. Poison ivy causes serious problems for most people, but a few lucky ones seem immune. Factors such as age, weight, and physical condition must be considered when evaluating poisoning incidents. Symptoms vary from source to source. In this text, they are representative of information given in literature cited. No attempt is made to offer treatments. Once a plant and its toxins are identified, the physician will be able to follow standard or preferred courses of action. Most scientific nomenclature in the text follows that in the *National List of Scientific Names* (USDA SCS, 1982). Ornamental species not included in the national list follow that of *Hortus Third* by L. H. Bailey (1976). The few exceptions not included in the national list or *Hortus Third* were taken from literature cited.

Acknowledgments

The authors would like to extend sincere appreciation to the following people who assisted in completion of this work:

To Rebecca Marks Westbrooks, Biology Teacher, for drawing the pictorial glossary.

To Dr. Brooks C. Metts and Mrs. Gwyndolyn Ballentine of the Palmetto Poison Center for technical assistance with toxicology and case histories and writing the Foreword.

To Dr. Wade T. Batson, Professor Emeritus, University of South Carolina, for sustained encouragement and writing the Preface.

To Mrs. Carol Winnberry of the Thomas Cooper Library, University of South Carolina, for valuable assistance in literature research.

To Leo Ehnis, PPQ Officer; Dr. Robert E. Eplee, Agronomist, USDA APHIS PPQ; and Earl Mullins, Inspector, U.S. Customs Service, for inspiriting patronage.

To Dr. Ritchie Bell, Department of Biology, University of North Carolina, for kind assistance in locating plant slides.

To Ken Scott, Robin Sumner and Earle Jackson of the University of South Carolina Press for efficacious support and continued patience.

To Dr. Julia Morton, University of Miami, for valuable inspiration.

Plant Toxins

Poisonous plants generally cause one or more of the following effects:
1. Internal poisoning and/or irritation.
2. Skin rash or dermatitis caused by allergenic or irritant compounds.
3. Skin photosensitization resulting in rash or dermatitis and possibly scars.
4. Airborne induced allergic reactions such as hayfever from pollen or respiratory irritation caused by volatile emanations from blossoms or foliage.

Poisonous principles generally fall into one of the following categories:

ALKALOIDS: complex, physiologically active, nitrogenous compounds that taste bitter and are usually insoluble in water. Thousands have been described, but not all are toxic. The same or similar compounds occur in closely related species, but occasionally one will appear in unrelated plants such as pyridine in poison hemlock (*Conium*) and tobacco (*Nicotiana*). Climate and water supply do not affect the alkaloid concentration of a plant very much.

GLYCOSIDES: compounds that yield one or more sugars (glycones) and one or more toxic aglycones. In pure form they are usually colorless, bitter crystalline solids; not all are toxic. Saponic glycosides form sapogenins (their aglycones) upon hydrolysis and suds when mixed with water. These compounds destroy red blood cells if absorbed into the blood. English ivy (*Hedera helix* L.) contains a saponin. Cyanogenic glycosides yield hydrogen cyanide as one product of hydrolysis. Amygdalin occurs in the seeds of apples, apricots, peaches and other members of the rose family (Rosaceae). Anthraquinone glycosides [e.g., in buckthorn (*Rhamnus*)] and their aglycones cause purging. Cardiac glycosides (cardenolides) occur in widely differing families such as the Apocynaceae (dogbone) and Ranunculaceae (crowfoot). Environmental factors, plant part and age all affect the glycoside content.

OXALATES: soluble oxalates occur in many plants but only a few contain toxic concentrations. Oxalic acid is one that is corrosive to animal tissues. Many plants of the arum family (Araceae) contain small crystals of insoluble calcium oxalate which cause intense oral irritation when plant parts are chewed.

PHYTOTOXINS (TOXALBUMINS): toxic protein molecules in a few plants that are similar to bacterial toxins in structure and reaction. Abrin occurs in rosary pea (*Abrus*) and ricin in castor bean (*Ricinus*).

POLYPEPTIDES and AMINES: nitrogenous compounds such as phenylethylamine and tyramine in mistletoe (Phoradendron).

Plant Toxins

RESINS: widely different compounds chemically, but share certain physical characteristics. At room temperature they are solid or semi-solid and brittle. They melt or burn easily, are soluble in organic solvents, insoluble in water and don't contain nitrogen. Chinaberry tree (*Melia*) contains resins.

MINERAL POISONINGS: may occur when plant species near industrial sites, along roadsides or on streambanks become toxic by accumulating minerals such as lead, copper, and arsenic.

PHOTOSENSITIZING COMPOUNDS: in PERCUTANEOUS PHOTOSENSITIZATION, moist skin comes in contact with certain plants. Subsequent exposure to sunlight causes symptoms similar to severe sunburn with itching and burning. Blisters appear later and may result in persisting scars. The compounds responsible for these reactions are psoralens which are furocoumarins composed of furan and coumarin rings. Plants in the fig family (Moraceae), parsley family (Apiaceae), rue family (Rutaceae) and the legume family (Leguminoseae) contain them. There has been much interest and concern in the use of psoralens in tanning lotions. PRIMARY and HEPATOGENIC PHOTOSENSITIZATION usually occur in animals. In the first, a toxic plant substance is consumed and absorbed unchanged into the bloodstream where it soon reaches the skin and is activated by sunlight to cause a toxic reaction. The red pigment hypericin in St. Johnswort (*Hypericum perforatum* L.) and fagopyrin in buckwheat (*Fagopyrum sagittatum* Gilib.) are examples of this. Psoralens also cause this type of reaction when ingested by animals and there is concern about humans eating foods that contain them as well. In secondary or hepatogenic photosensitization, the compound is phylloerythrin, an anaerobic fermentation product of chlorophyll made by bacteria in the stomach of ruminants. When absorbed from the gut, this compound is normally removed from the bloodstream by the liver and excreted in bile before it reaches the skin. However, if the liver is damaged or otherwise prevented from excreting the toxin, it will reach the skin and cause a phototoxic reaction. Lantadene in *Lantana camara* causes this syndrome in livestock.

Pictorial Glossary

Inflorescences

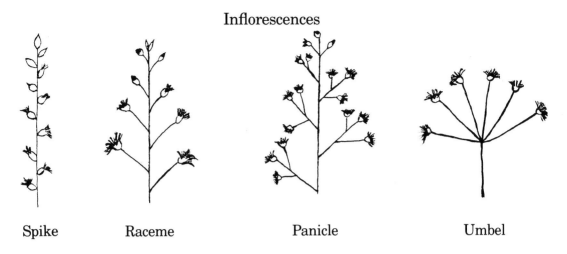

Spike   Raceme   Panicle   Umbel

Cyme (Compound Umbel)   Corymb

Parts Of A Flower

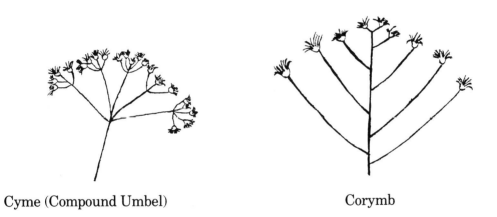

Leaf Shapes

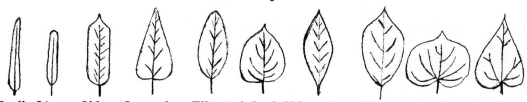

Needle Linear Oblong Lanceolate Elliptical Oval Oblanceolate Obovate Reniform Cordate

Leaf Arrangement

Alternate　　　　　　Opposite　　　　　　Whorled

Leaf Composition

Simple Leaf　　　Pinnately Compound Leaf　　　Bi-Pinnately Compound Leaf

Palmately Compound Leaf　　　Palmate-trifoliate

False Sago Palm
Cycas circinalis L.
Cycadaceae

DESCRIPTION: Palm-like plant with a thick rough trunk; to 4.5 m tall.
LEAVES: Fern-like, pinnately divided, 1.5–2.4 m long; leaflet pairs many.
FRUIT: Male cones erect, to 30 cm tall, woolly, red to yellow; female cones rounded; seeds round to oval, to 4 cm long, orange-yellow when ripe; with a large white kernel in a thin shell.
OCCURRENCE: Cultivated in coastal areas; North Carolina and south.
TOXICITY: The glycoside cycasin, its aglycone methyl-azoxymethanol (114, 149) and α-amino-β-methyl-aminoproprionic acid (83) occur in the seed kernel, roots and trunk pith.
SYMPTOMS: Headache, vomiting, vertigo, swelling of the stomach, diarrhea, jaundice, anemia, muscle paralysis, hemorrhage and death (106, 114, 175); dermatitis (142).
NOTES: Medicinally, the prepared starch, the fresh exudate and the grated fresh seeds or stems have been recommended as an external remedy for snake bite, swelling, wounds and boils. In Guam, the grated fresh nuts of "fadang," as it is locally known, are used by natives to treat tropical leg ulcers (175). In some tropical areas, the nut is cut into small

pieces, soaked in water and dried. The dried pieces are then ground into powder and used as an edible starch (114) as are the pith and roots (142). The uncooked fruit husk has been eaten green or dried in Guam and the fruit with sugar in India. The shoots, leaves and ripe seeds are also boiled as a vegetable in Java, Sri Lanka, Sumatra and Fiji (175). Toxicity is usually associated with improper washing, but frequent or long-term use of even properly prepared material may cause harmful effects (114) such as liver damage and cancer (142). While exploring Queensland and northern Australia in 1770, a number of Captain Cook's men experienced severe vomiting from eating cycad nuts. They concluded that local natives used the plant for food after observing shells from the nuts around campfire sites (175). The male cone is said to emit a noxious odor that is a respiratory irritant to some people (142).

Coontie, Florida Arrowroot

Zamia spp.
Cycadaceae

DESCRIPTION: Fern-like plants with thick, starchy, underground stems.
LEAVES: Pinnately divided with about 15 pairs of leaflets.
CONES: Staminate cones slender, 7–12 cm long; pistillate cones on separate plants, 12–17 cm long with orange-coated seeds.
OCCURRENCE: Woods and borders, in coastal areas; Georgia, Florida and Alabama; also cultivated.
TOXICITY: The glycoside cycasin, its aglycone methyl-azoxymethanol and α-amino-β-methyl-aminoproprionic acid occur in the seed kernel and raw roots (83).
SYMPTOMS: Headache, vomiting, gastroenteritis, jaundice, anemia, hemorrhage, diarrhea, paralysis and death (106, 114).
NOTES: In 1539, one of DeSoto's men was fatally poisoned by the raw root while in Florida (142), but it was an important starch in the diet of Indians and early settlers. They peeled, chopped and repeatedly washed it to remove the toxins (128). The root is also chewed in Guatemala to relieve cough and improve one's singing voice. Cycad starch has been produced and/or marketed in Guam, Japan, the Dominican Republic, the United States and other areas. It was produced in southern Florida as early as 1845. At the peak of production, 10–15 tons of *Zamia* root were processed weekly by mills along the Miami River. The starch marketed under the name of Florida arrowroot was used for baby food, in chocolates, biscuits and spaghetti (175). *Z. pumila* L. is wild and cultivated in Florida.

Ginkgo
Ginkgo biloba L.
Ginkgoaceae

DESCRIPTION: Trees with irregular branching patterns.
LEAVES: Deciduous, 2-lobed, fan-shaped, on short branchlets; veins prominent, somewhat parallel.
CONES: Staminate small, slender, many in early spring; pistillate on separate trees, solitary or paired.
FRUIT: Globular drupes, yellow-orange; pulp foul smelling when ripe and split open in Autumn; seeds smooth, cream-colored.
OCCURRENCE: Ginkgo survived the last ice age in China and is the last survivor of the family Ginkgoaceae; widely cultivated in the U.S.
TOXICITY: The alkyl resorcinol 6(pentadec-8-enyl)-2, 4-dihydroxybenzoic acid (60), an alkyl phenol and ginkgolic acid (124) are found in the fruit pulp and seeds.
SYMPTOMS: The fruit juice may cause severe skin irritation (72, 106, 138, 142, 165); internally it irritates the stomach, intestines and kidneys (90, 138, 142).
NOTES: In one case, a man who ate 2 pieces of fruit pulp experienced swelling, tingling, burning and soreness of the lips, mouth and throat (20). In Japan, the leaf has been used since ancient times to protect books from insects by inserting it between the pages (171).

Juniper, Red Cedar
Juniperus spp.
Cupressaceae

DESCRIPTION: Evergreen, resinous shrubs or trees; bark thin.

LEAVES: Needle- or scale-like.

CONES: Unisexual, usually on different trees; staminate small, cylindric, at ends of branchlets; pistillate small, leathery, globular, also at branchlet ends.

FRUIT: Fleshy to leathery, berry-like cones, ripening in 1–3 seasons.

OCCURRENCE: Native and cultivated; widely spread.

TOXICITY: Volatile oils which include thujone (106).

SYMPTOMS: Large doses may cause purging, but repeated small amounts taken herbally may result in kidney damage (106). The oils have been reported to have irritant and allergenic properties, but recent dermatological data have them to be generally nontoxic externally (108).

NOTES: Hopi Indians boiled the leaves for stomachache, earache, sore throat and constipation (116). The berries and young twigs have been used for digestive problems due to underproduction of stomach acid (113). Bronchitis has been treated in Appalachia by inhaling the steam from a boiling mixture of the fruit and foliage (102). Oils from eastern red cedar (*J. virginiana* L.) have been used for abortions and sometimes proved fatal (5).

Yew, Ground Hemlock

Taxus spp.
Taxaceae

DESCRIPTION: Evergreen shrubs or small trees; bark reddish-brown, scaly.
LEAVES: Linear, flattened, sometimes curved, spirally arranged.
CONES: Staminate globe-like with overlapping scales and 4–8 pollen sacs; pistillate usually on separate trees with a single ovule and a circular disk around the base.
FRUIT: A small, hard-shelled seed surrounded by a fleshy red aril.
OCCURRENCE: Native and introduced; Florida yew (*T. floridana* Nutt.) is a small tree or large shrub in moist, shady ravines in Gadsden and Liberty Counties in NW Florida; English yew (*T. baccata* L.) and Japanese yew (*T. cuspidata* Sieb. & Zucc.) are cultivars.
TOXICITY: The bark, foliage and seeds, green or dry contains alkaloids taxine A and B (93, 106, 109).
SYMPTOMS: Pupil dilation, delirium, difficult breathing, abdominal pain, vomiting, diarrhea, weakness, convulsions and slow heartbeat; large amounts may cause death (72, 106, 134).
NOTES: Yew has always been associated with witchcraft, mourning and churchyards (103). In one case of poisoning, a young man took a decoction of yew foliage about 2 A.M. one morning to evaluate its effects before giving it to his girlfriend as an abortifacient. Shortly before 5, his brother, with whom he shared a room, awakened and found him unconscious with rattling respirations. He was seen at a clinic in a deep coma at 5:25. Despite treatment with circulatory drugs, he died about 30 minutes later of respiratory failure (58). The berry-like aril around the seed is reported as edible in small amounts (72).

Japanese Yew
Podocarpus macrophylla (Thunb.) D. Don
Podocarpaceae

DESCRIPTION: Evergreen tree; to 15 m tall.
LEAVES: Alternate, crowded, 5–12 cm long, blade dark green, midrib lighter, lanceolate.
FRUIT: Light blue berries.
OCCURRENCE: Native to Japan; widely cultivated.
TOXICITY: Leaves and berries contain unidentified toxins.
SYMPTOMS: Gastrointestinal irritation, nausea, vomiting and diarrhea (49).
NOTES: The berries are sometimes mistaken for grapes by children.

Fishtail Palm

Caryota mitis Lour.
Arecaceae

DESCRIPTION: Multi-stemmed palm, forming clumps; 7–10 m tall.
LEAVES: With many small, wedge-shaped sections; ends ragged like fish tails; to 2.7 m long.
FLOWERS: Small, in long clusters.
FRUIT: Round, dark red when ripe, in many-stranded bunches like strings of beads.
OCCURRENCE: Native to Asia, now cultivated in south Florida as an ornamental that continuously blooms and fruits.
TOXICITY: Fruit pulp and juice with stinging crystals (128).
SYMPTOMS: Painful skin irritation with itching for several hours (72, 127, 128); internal irritation occurs if consumed (142, 165).
NOTES: The fruit kernels are reported as edible (127, 128, 142).

Jack-in-the-Pulpit
Arisaema triphyllum
(L.) Schott
Araceae

DESCRIPTION: Tuberous perennial herb.

LEAVES: Trifoliate, 8–20 cm long.

FLOWERS: On a fleshy spike enclosed by a yellow to purple spathe; pistillate at bottom, staminate in middle; stamens 2–5; perianth none; March-early July.

FRUIT: Berry, 1–3 seeded.

OCCURRENCE: Rich woods; throughout the U.S. and Canada; sometimes cultivated.

TOXICITY: Needle-like crystals of calcium oxalate and possibly other toxins as well.

SYMPTOMS: When plant parts are chewed, crystals become embedded in the membranes of the mouth and throat causing intense burning and irritation. Nausea, vomiting and diarrhea may also occur. Exposure to the sap may cause skin dermatitis.

NOTES: The rootstock has been used for asthma, whooping cough and rheumatism (113); raw roots have been used with lard in a salve for ringworm and as a counterirritant to snakebite (5). Pawnee Indians sprinkled the dried and powdered roots on the head and temples for headache (109). Other Indians ate them after heating or drying to remove the toxins (102). In one case of poisoning, 2 teenage girls suffered intense mouth and throat irritation after being given pieces of the plant to chew. Poisonings also occur in the fall when the ripe berries are eaten by children (89).

Caladium

Caladium spp.
Araceae

DESCRIPTION: Tuberous, perennial herbs.
LEAVES: Simple, variegated.
FLOWERS: With a yellow-green spathe enclosing a spadix.
OCCURRENCE: Introduced from the tropics; a border plant in the South, potted elsewhere.
TOXICITY: Needle-like crystals of calcium oxalate and possibly other toxins as well.
SYMPTOMS: When plant parts are chewed, crystals become embedded in the membranes of the mouth and throat causing intense burning and irritation. Nausea, vomiting and diarrhea may also occur. Exposure to the sap may cause skin dermatitis.
NOTES: The cooked leaves and roots are eaten in tropical America; the powdered leaves are used as an insecticide in the Phillipines (128).

Dumbcane
Dieffenbachia spp.
Araceae

DESCRIPTION: Tall, perennial herbs.
LEAVES: Large, simple, smooth, mottled with different shades of green.
OCCURRENCE: Introduced from the tropics; a cultivated pot plant with many varieties.
TOXICITY: Needle-like crystals of calcium oxalate and possibly other toxins as well.
SYMPTOMS: When plant parts are chewed, crystals become embedded in the membranes of the mouth and throat causing intense burning and irritation. Nausea, vomiting and diarrhea may also occur. Exposure to the sap may cause skin dermatitis.
NOTES: In one case, a woman bit into a stalk of dumbcane. Intense pain soon developed, so she spat out the pulp and juices before swallowing any. Six hours later she was taken to an emergency room with severe pain and swelling of the face, tongue and palate. Blisters had formed on the tongue and mucosa, salivation was excessive, she could not swallow and speech was unclear. After the pain was relieved by a parenteral narcotic, she was able to swallow a demulcent. By the next day, she could swallow soft foods and liquids, but the swelling and inflammation remained. Three days after the incident the swelling was less, but the pain was still severe and necrosis was beginning on the left side of the tongue and cheek. By the 13th day, the facial configuration had returned to normal (46).

Hunter's Robe
Epipremum aureum Bunt.
Araceae

SYNOMYMS: *Raphidiophora aurea* Birdsey; *Pothos aureus* Lind. & Andre; *Scindapsus aureus* Engl.
DESCRIPTION: High climbing vine; stems rope-like with aerial roots.
LEAVES: Heart-shaped, to 5 dm long, yellow-streaked.
FLOWERS: Rare in cultivation.
TOXICITY: Needle-like crystals of calcium oxalate and possibly other toxins as well.
SYMPTOMS: When plant parts are chewed, crystals become embedded in the membranes of the mouth and throat causing intense burning and irritation. Nausea, vomiting and diarrhea may also occur. Exposure to the sap may cause skin dermatitis.
OCCURRENCE: Usually growing on tree trunks or as pot plants in the southern U.S.
NOTES: Nurserymen of south Florida term irritation from the clear watery juice "pothos poisoning" (128).

Split Leaf Philodendron

Monstera deliciosa Liebm.
Araceae

DESCRIPTION: Woody-stemmed, climbing vine.
LEAVES: Leathery, heart-shaped with prominant lobes and irregularly located, oblong holes throughout the blade.
FLOWERS: Small, on a spadix with a green spathe.
FRUIT: One foot or more long, with a green rind and white flesh.
OCCURRENCE: Introduced from the tropics; grown as an ornamental in south Florida and as a pot plant farther north.
TOXICITY: Needle-like crystals of calcium oxalate and possibly other toxins as well.
SYMPTOMS: When plant parts are chewed, crystals become embedded in the membranes of the mouth and throat causing intense burning and irritation. Nausea, vomiting and diarrhea may also occur. Exposure to the sap may cause skin dermatitis.
NOTES: The juice may irritate sensitive skin and eyes. The fruit is reportedly edible, but may cause allergic reactions in some people (142).

Golden Club
Orontium aquaticum L.
Araceae

DESCRIPTION: Rhizomatous aquatic herb from 2–6 dm tall.
LEAVES: In a basal cluster, ovate to elliptical, non-wettable.
FLOWERS: Perianth parts 4–6, hooded, regular, perfect, on a long spadix; stamens 4 or 6.
FRUIT: Bladdery, blue-green with 1 globose seed.
OCCURRENCE: Swamps, lake shores; throughout most of the East; sometimes cultivated.
TOXICITY: Needle-like crystals of calcium oxalate and possibly other toxins as well.
SYMPTOMS: When plant parts are chewed, crystals become embedded in the membranes of the mouth and throat causing intense burning and irritation. Nausea, vomiting and diarrhea may also occur. Exposure to the sap may cause skin dermatitis.

Philodendron
Philodendron spp.
Araceae

DESCRIPTION: Climbing vines or shrubs.

LEAVES: Large, variously shaped, simple, glossy, dark green.

OCCURRENCE: Native to the tropics; cultivated in the southern U.S. and as a pot plant in other areas, for the lush green foliage.

TOXICITY: Needle-like crystals of calcium oxalate and possibly other toxins as well.

SYMPTOMS: When plant parts are chewed, crystals become embedded in the membranes of the mouth and throat causing intense burning and irritation. Nausea, vomiting and diarrhea may also occur. Exposure to the sap may cause skin dermatitis.

NOTES: The incidence of dermatitis from these plants is not great, but several cases have been recorded where they were handled at work or home. In one case, a 32-year-old woman who was a secretary at a large seed company reported eruptions on her right hand on several occasions for four or five years. Careful questioning revealed that she had *P. cordatum* vines at the office which she watered and pruned. Patch tests confirmed her susceptibility to philodendron leaves. The eruptions were erythematous and slightly vesicular (12). These popular house and patio plants should be safeguarded especially from children to avoid poisonings.

Water Lettuce
Pistia stratiotes L.
Araceae

DESCRIPTION: Floating, stemless herb; roots submerged, hollow, feathery.

LEAVES: Spongy, wedge-shaped, 2–12 cm long, several-veined, base inflated, thick.

FLOWERS: Small, on a short spadix which is inside a hairy, tubular spathe; staminate above, pistillate below.

FRUIT: Green, slimy inside; seeds many, brown.

OCCURRENCE: Forming colonies in ditches, ponds or pools; often spreading by runners; native from Florida to Texas.

TOXICITY: Needle-like crystals of calcium oxalate and possibly other toxins as well.

SYMPTOMS: When plant parts are chewed, crystals become embedded in the membranes of the mouth and throat causing intense burning and irritation. Nausea, vomiting and diarrhea may also occur. Exposure to the sap may cause skin dermatitis.

NOTES: Aquaria with this plant should be safeguarded to prevent accidental poisoning in children.

Skunk Cabbage

Symplocarpus foetidus (L.) **Salisb.**

Araceae

DESCRIPTION: Foul-smelling, perennial herb; rhizome thick.

LEAVES: Cordate, with netted veins.

FLOWERS: Perianth parts 4; stamens 4; ovary 1-celled on a fleshy spadix surrounded by a purple, striped or spotted spathe; Spring.

FRUIT: Globular, fleshy spadix around spherical seeds; remaining on the ground well after the leaves disappear in the Fall.

OCCURRENCE: Swamps, wet woods: southern Canada, south in the mountains to Georgia; may be cultivated.

TOXICITY: Needle-like crystals of calcium oxalate and possibly other toxins as well.

SYMPTOMS: When plant parts are chewed, crystals become embedded in the membranes of the mouth and throat causing intense burning and irritation. Nausea, vomiting and diarrhea may also occur. Exposure to the sap may cause skin dermatitis.

NOTES: The roots have been used for respiratory ailments, dropsy, rheumatism and nervous disorders (113). Overdoses may cause nausea, vomiting, vertigo, disturbed vision and headaches (109). Micmac Indians crushed the leaves and inhaled the pungent oils to cure headaches (109). The dried roots mixed with honey has been taken for asthma and chest ailments (102).

Oyster Plant

Rhoeo spathacea (Swartz) Stearn
Commelinaceae

DESCRIPTION: Herb with succulent stems, to 4.5 dm tall.
LEAVES: Flat, stiff, to 3.5 dm long; dark green above, purple to green beneath.
FLOWERS: White, small, between 2 purple bracts; with 3 petals.
FRUIT: Capsules with 2–3 valves; seeds slender.
OCCURRENCE: Native to tropical America; a common ground cover in south Florida and other warm areas; potted elsewhere.
TOXICITY: The watery juice contains unidentified toxins.
SYMPTOMS: Chewing the leaves or stems causes burning of the mouth and throat, abdominal pain and gastroenteric irritation; the juice causes respiratory irritation if inhaled (128, 142).
NOTES: The plant has been used in folk medicine of the Caribbean and South America as an astringent, a diuretic and a cough remedy (128).

Purple Queen
Tradescantia pallida Hunt
Commelinaceae

SYNONYM: *Setcreasea purpurea* Boom.
DESCRIPTION: Trailing, reclining or climbing herb; stems succulent, easily broken.
LEAVES: Purple, to 18 cm long.
FLOWERS: Solitary, regular, pink-lavender, enclosed by two bracts.
OCCURRENCE: Native to Mexico; introduced as a ground cover in Florida; north to South Carolina and on the Gulf Coast; otherwise cultivated.
TOXICITY: The sap contains unidentified toxins.
SYMPTOMS: Eye inflammation and skin dermatitis (72, 106, 109, 127, 128, 142, 165).
NOTES: Mexican women reportedly sometimes use the sap to inflame and increase the color of their cheeks (128).

Asparagus

Asparagus officinalis **L.**
Liliaceae

DESCRIPTION: Erect, glabrous, perennial herbs; stems several, 8–20 dm tall, from a crown.

LEAVES: Stem leaves alternate, scale-like, to 1 cm long; branch leaves very small 5–27 mm long, 1–7 in a fascicle.

FLOWERS: Pendent, 1–3 in axils; pedicels jointed; perianth 6-parted, bell-shaped, yellow-green; stamens included; April-June.

FRUIT: Red, globose berries, 5–10 mm in diameter, with 3–6 seeds.

OCCURRENCE: Native to Old World north temperate areas; grown except in very hot areas commercially and in gardens; sometimes escaping to waste ground and thickets.

TOXICITY: Raw shoots and berries contain unidentified toxins.

SYMPTOMS: Consuming the young raw shoots may cause dermatitis with reddened skin to painful swelling and blisters (90); allergic contact dermatitis (90, 165).

NOTES: The cooked young shoots have been used for food since ancient Rome and Greece (90). Asparagus fern (*A. sprengeri* Regel) may also be toxic if chewed or eaten (142).

Autumn Crocus, Meadow Saffron

Colchium autumnale L.
Liliaceae

DESCRIPTION: Bulbous herb.
LEAVES: Long, basal.
FLOWERS: Purple, white or yellow; 1–6; appearing in the Fall after the leaves have withered.
OCCURRENCE: Introduced and cultivated in gardens of North America; sometimes naturalizing in damp woods and fields.
TOXICITY: The alkaloid colchicine and related compounds are concentrated in the flowers and bulbs (55, 72, 97, 106, 109, 142).
SYMPTOMS: In 2–7 hours: thirst, nausea, burning of the mouth, throat and stomach, difficulty in swallowing, abdominal pains, weakness, vomiting, bloody diarrhea, labored breathing, kidney damage, dehydration, shock, collapse, accelerated pulse and occasional coma; recurring symptoms may result in respiratory failure and death; when recoveries occur, there may be temporary hair loss (72, 101, 109, 128); the leaves may cause dermatitis (55, 129, 165).
NOTES: Toxic levels of the alkaloids may accumulate in the body of animals. In lactating animals, one pathway of excretion is through the milk; therefore, contaminated milk from a family cow may cause poisoning (4). Fatalities have occurred following administration of the plant extract for gout and rheumatism; by children who eat the flowers, and by persons eating the bulbs after mistaking them for onions (109). In one case of poisoning, a 16-year-old girl who had frequently threatened suicide, ate more than a dozen of the flowers on September 3, 1967. After a few hours, she had profuse diarrhea and vomiting that

persisted through the night. The next morning, she was admitted to a hospital with shock, intense thirst, dehydration and abdominal pain. Despite therapy, which included atropine, vomiting and diarrhea continued. Her condition worsened until she died soon after 10 P.M. It was estimated that she had ingested about 270 mg (51).

Lily-of-the-Valley
Convallaria majalis L.
Liliaceae

DESCRIPTION: Fragrant, perennial herb with creeping rootstock.

LEAVES: Simple, sheathing, ovate, to 25 cm long.

FLOWERS: On leafless stalks, in 1-sided racemes, white, bell-shaped, drooping.

FRUIT: Red-orange berry, 1–3 seeded.

OCCURRENCE: Native to Britain and Europe; cultivated in shady gardens throughout the U.S.; sometimes escaping to roadsides, thickets and open woods.

TOXICITY: All parts contain the cardiac glycosides convallarin, convallamarin and convallatoxin (49, 72, 93, 101); the aglycone of convallatoxin is convallatoxigenin (101).

SYMPTOMS: Burning of the mouth and throat, nausea, vomiting, abdominal pain, purging, dilated pupils, decreased pulse rate, irregular heartbeat, cold clammy skin, collapse, coma, circulatory failure and death (4, 49, 134).

NOTES: One medieval legend associates this plant with St. Leonard, a favorite of King Clovis of France. In doing battle with the devil in the form of a dragon to decide who should live in and rule the forest, Leonard's blood produced this flower wherever it was spilled. The dragon's blood produced poisonous weeds wherever it touched (116). The plant has been used since ancient Greece for heart disease, dropsy and as a purgative (93). Children are attracted to the bright flowers and berries, and have been killed drinking water from a vase containing the flowers (4, 90). The plant has also been used as an African arrow poison (171).

Glory Lily
Gloriosa superba L.
Liliaceae

DESCRIPTION: Slender herb or vine with tuberous roots.

LEAVES: Alternate or seeming opposite, 10–18 cm long, simple; tips tendril-like.

FLOWERS: Red or yellow on red stalks; sepals 6, lobe edges crinkled; petals 6; pistil and 6 stamens projecting downward.

FRUIT: Capsules, 5–7 cm long.

OCCURRENCE: Native to Old World tropics; cultivated in Florida and potted elsewhere.

TOXICITY: All plant parts contain colchicine and superbine; the flowers contain lumi-colchicine (109, 128).

SYMPTOMS: In 2–7 hours: thirst, nausea, burning of the mouth, throat and stomach, difficulty in swallowing, abdominal pains, weakness, vomiting, bloody diarrhea, labored breathing, kidney damage, dehydration, shock, collapse, accelerated pulse and occasional coma; recurring symptoms may result in respiratory failure and death; when recoveries occur, there may be temporary hair loss (72, 101, 109, 128).

NOTES: The tubers have been used for murder and suicide in Burma and India (109, 128). Deaths have also occurred from use of the plant in African and Asian folk medicine (128). On September 16, 1964, a 21-year-old woman in Sri Lanka became ill after an evening meal of boiled yams that included glory lily. She ingested about 125 g of the tubers which were estimated to contain about 350 mg of colchicine. Two hours later, she began vomiting and 8 hours later had diarrhea, both of which persisted through the night. She was admitted to a hospital the next day collapsed, restless, dehydrated and unconscious. Her condition improved, but a relapse occurred the next day. On the 19th, she improved once again, but was found to have a sub-conjunctival hemorrhage of the left eye on the 22nd. Her menstrual period, which was ending on the 16th, continued for a further 20 days. Twelve days after admission, her hair began falling out and was gone within 2 days. She was discharged on the 16th day, but was bald by the 23rd day. After 5 months, her hair had regrown to a length of 5–8 cm (63).

Hyacinth
Hyacinthus orientalis L.
Liliaceae

DESCRIPTION: Bulbous perennial herb.
LEAVES: To 30 cm long and 2 cm wide.
FLOWERS: Variously colored in a cluster; on a stalk.
FRUIT: Capsules, with 3 cells; seeds black.
OCCURRENCE: Introduced from the Mediterranean area; cultivated in gardens, lawns and pots.
TOXICITY: The bulbs contain crystals of calcium oxalate (55, 109).
SYMPTOMS: Stomach cramps, diarrhea and vomiting (72, 142); allergic contact dermatitis occurs in some people (49, 72, 106, 142, 165).
NOTES: According to legend, the first flowers with this name sprang from the blood of Hyacinthus, a fatally wounded Greek youth, whom Apollo immortalized by naming the flower for him (116).

Star-of-Bethlehem

Ornithogalum umbellatum L.
Liliaceae

DESCRIPTION: Glabrous, perennial, bulbous herb; stems leafless, to 31 cm.

LEAVES: Basal, linear, midrib light green; usually withering at the time of flowering.

FLOWERS: Perianth white, persisting, star-like, 6-parted, each with a green stripe on the back, corymbose; March-June.

FRUIT: Several-seeded, 3-lobed capsule.

OCCURRENCE: Native to the Mediterranean area; naturalized in fields, waste places, along roadsides and lawns of the eastern and central U.S. and Canada; may be cultivated.

TOXICITY: The bulbs contain convallatoxin and convalloside (109).

SYMPTOMS: Gastrointestinal upset with nausea (109).

NOTES: According to legend, this was among the grasses and herbs on which the Christ child lay in the manger. After it blossomed and formed a wreath around his head, he blessed it and called it the Star-of-Bethlehem. It is customary in Italy to deck the manger with these flowers at Christmas in remembrance of the legend (116). Children have been poisoned by the flowers and bulbs (72, 109).

False Hellebore
Veratrum viride Ait.
Liliaceae

DESCRIPTION: Herbaceous perennial, 9–24 dm tall; rootstock thick, vertical; stems pubescent, leafy.

LEAVES: Alternate, oval, 3-ranked up the stem, 15–30 cm long, base sheathing the stem.

FLOWERS: In large terminal panicles, greenish-yellow; perianth 6-parted, hairy; May-August.

FRUIT: Capsules with yellowish-brown seeds.

OCCURRENCE: Native in wet areas, woodlands, on streambanks; west to Minnesota south in the mountains to Georgia.

TOXICITY: The alkaloids veratrine, jervine, cevadine and others occur in European species and may occur in American species as well (129). The glyco-alkaloid veratrosine is also reported (101).

SYMPTOMS: Burning of the mouth and throat, excess salivation, headache, vomiting, diarrhea, stomach pain, weakness, sweating, spasms and general paralysis; other symptoms may include hallucinations, shallow breathing, irregular pulse, low temperature, collapse, convulsions, asphyxia and death (72, 101, 142, 165).

NOTES: Deaths are probably rare due to vomiting which removes the slowly absorbed toxins (142). All authorities through the centuries have considered this plant as poisonous, but most cases occur through misuse of medicinal preparations from it (109). Cherokee Indians used a decoction from it to relieve body pains (109). Colonists used it in treating shingles and head lice (102). On May 6, 1953, the cook of a 14-man labor party of the Korean Service Corps picked several different kinds of wild greens and made them into a soup which was eaten at noon.

False Hellebore

At 1 P.M., 3 of the men began vomiting and by 3 P.M. were seeing yellow and green spots, circles and yellow sheets of flame before the eyes. At this time, 6 more men became ill and the remaining men reported similar symptoms over the next few hours. A U.S. Army Battalion surgeon was requested to observe and advise treatment for the patients who were vomiting, had pulse rates below 40, injected conjunctivas, swollen abdomens and difficulty in opening the eyes widely. Complete recovery came within 15 hours of treatment with sodium chloride solution and epinephrine. When samples of the greens were identified, only *V. nigrum* L. var. *japonicum* Baker had toxic principles (131).

Black Snakeroot, Death Camas

Zigadenus densus (Desr.) Fernald

Liliaceae

DESCRIPTION: Perennial herb; rootstock thick and horizontal; stems smooth, leafy, to 9 dm tall.

LEAVES: Mostly basal, narrow, sessile.

FLOWERS: In terminal panicles; white or cream-colored; perianth parts 6, with 1–2 yellowish glands on the upper side at the base; April-June.

FRUIT: Conical, 3-celled capsule.

OCCURRENCE: Wet, open woods and swamps; Coastal Plain, Virginia and south.

TOXICITY: The alkaloids zygadenine, zygacine and others occur mostly in the roots (72, 109, 142).

SYMPTOMS: Burning in the mouth and throat, excessive salivation, weakness, abdominal pain, gastro-intestinal irritation, vomiting, diarrhea, low temperature, slow and weak heartbeat, low blood pressure, coma and death (142).

NOTES: Early pioneers were poisoned when they confused these bulbs with those of similar plants such as wild onion (*Allium*) or edible camas (*Camassia*). Rural families sometimes boiled edible camas bulbs or steamed them between layers of fern leaves over an open fire. In one case, a 2.5-year-old boy was poisoned by death camas bulbs that had been roasted by several children over a bonfire. Within an hour, he began to stagger, vomited several times, fell unconscious and was then taken to a hospital. With treatment, he regained consciousness after 3 hours and was discharged the next day with only mild diarrhea (31). Members of the Lewis and Clark expedition experienced violent intestinal upset from flour made with zigadenus bulbs (10).

Century Plant

Agave spp.
Amaryllidaceae

DESCRIPTION: Perennial herbs; roots thick, fibrous.
LEAVES: Clustered in a rosette, thick, fleshy, often spiny, rising upward and outward.
FLOWERS: Perianth tubular, 6-parted, scapose; stamens 6.
FRUIT: Leathery capsules with flattened seeds.
OCCURRENCE: Introduced; cultivated North Carolina and south; sometimes escaping.
TOXICITY: The sap contains oxalic acid, lycorine and a saponin (106, 128).
SYMPTOMS: Itching and skin burning with red welts and eruptions that may last for days (72, 128).
NOTES: Agave has been called the "Mexican tree of life," for it has provided them with food, fodder, dye, soap, twine and roofing (113). The sap has been used as a laxative and juice from the roots has been used in a scalp tonic (113). Trimming the lower leaves of an agave with a power mower has caused severe symptoms of blisters lasting almost a week (128).

Amaryllis

***Amaryllis* spp.**
Amaryllidaceae

DESCRIPTION: Bulbous herbs.
LEAVES: Many, strap-like.
FLOWERS: Large, fragrant, umbellate; perianth 6-parted; ovary 3-celled; corona inconspicuous.
OCCURRENCE: Introduced; cultivated.
TOXICITY: Bulbs and seeds contain alkaloids such as haemanthamine, hippeastrine, lycorine, tazzetine, amaryllidine and others (106, 128, 142).
SYMPTOMS: Other lycorine containing plants have caused gastroenteritis, vomiting, diarrhea and shivering (142).

Milk & Wine Lily
Crinum spp.
Amaryllidaceae

DESCRIPTION: Bulbous herbs.

LEAVES: Smooth, light-green, curving outward; 1 m or so long.

FLOWERS: Variously colored, in clusters on a long stalk; fragrant; funnel-shaped or with narrow curved segments; flowers sometimes develop into bulbs.

OCCURRENCE: Native and introduced; cultivated, some persisting; North Carolina and south.

TOXICITY: Alkaloids such as lycorine and crinidine are found in the bulbs (106, 128, 142).

SYMPTOMS: Raw bulbs cause gastroenteric irritation, chiefly of the gastric mucosa, vomiting and diarrhea (142).

Jonquil, Daffodil
Narcissus spp.
Amaryllidaceae

Daffodil

Jonquil

DESCRIPTION: Scapose, glabrous, perennial herbs with bulbs.
LEAVES: Linear or sword-shaped.
FLOWERS: Solitary or in umbels, subtended by a sheathing membranous spathe, which is partially open on one side; perianth tubular, petals and sepals spreading, white or yellow; stamens included or barely exserted from the perianth tube.
FRUIT: Capsules with many black seeds.
OCCURRENCE: Introduced ornamental; in yards, also potted; persisting for many years.
TOXICITY: Bulbs of *N. poeticus* L., *N. jonquilla* L. and *N. pseudonarcissus* L. contain alkaloids such as galanthamine, haemanthamine, lycorine and others (142, 163). Bulbs of all species contain crystals of calcium oxalate (55).
SYMPTOMS: Eaten in quantity, the bulbs cause gastroenteritis, nausea, vomiting, purging, trembling, convulsions and death (72, 101); bulbs, petals and sap cause allergic contact dermatitis (49, 106, 142, 165).
NOTES: Dermatitis from the bulbs is known as *lily rash*. The bulb has been crushed, mixed with honey and used externally for chronic joint pains (103). Confusion between these bulbs and ordinary onions has caused poisoning (4).

Iris, Flag
Iris spp.
Iridaceae

DESCRIPTION: Perennial herbs; stems erect from rhizomes.

LEAVES: Narrow, sword-shaped, long, some sheathing the stem.

FLOWERS: Large, blue or violet with yellow markings; sepals sometimes bearded with yellow or white trichomes.

FRUIT: Capsules.

OCCURRENCE: Native and introduced; woods, swamps, around homes; some cultivated.

TOXICITY: The rootstock contains irisin, an acrid resinous substance (106, 129, 142).

SYMPTOMS: In large amounts: gastroenteric irritation, chiefly to the intestinal mucosa, vomiting and purging (106, 129, 142); rhizomes and other parts may cause allergic contact dermatitis (49, 72, 90, 106, 142).

NOTES: In Greek mythology, Hera was so impressed by Iris, the goddess of the rainbow, that she commemorated her with a flower (48). Botanist William Bartram reported the use of *I. versicolor* L. in an Otasses Indian village in December 1777. Each village cultivated a patch of it for the highly cathartic roots (168). Indians also lightly boiled the roots, crushed and made them into a poultice for burns, bruises and sores (109).

Lady Slipper Orchid

***Cypripedium* spp.**
Orchidaceae

DESCRIPTION: Erect, rhizomatous herb with glandular hairs.
LEAVES: Basal or cauline, broad with conspicuous veins sheathing the stem.
FLOWERS: One-several, showy, with bracts; sepals 3, spreading, free or with 2 of them united under the sac-like lip; petals free, spreading, lip inflated and hollow.
FRUIT: Capsules.
OCCURRENCE: In damp or dry woods, swamps; Georgia and north; sometimes cultivated.
TOXICITY: Glandular hairs on the leaves and stems contain an irritant chemical (90).
SYMPTOMS: Allergic contact skin irritation may occur soon after contact, with severe irritation several hours later (72, 90, 106, 109, 129, 165). Possible internal irritation if consumed (138).
NOTES: *Cypripedium* is derived from the Greek words *kypris* meaning Venus and *podion,* meaning "slipper" or "little foot". Extracts from the plant were once used as sedatives to treat hysteria and neuralgia (109). The irritant nature of these plants was first suspected around 1875 when Professor H. H. Babcock of Chicago discovered that his recurring attacks of poison ivy were in fact caused by lady's slipper. The action of the irritant oil, which is abundant during the fruiting season, was later found to be similar to that of toxicodendrol in poison ivy (*Toxicodendron radicans*) (35).

Fig
Ficus carica L.
Moraceae

DESCRIPTION: Small to medium-sized tree or large shrub, to 10 m.
LEAVES: Large, deeply 3–5 lobed, with stiff hairs on the upper side.
FLOWERS: Staminate and pistillate produced inside a globular structure that becomes the fruit.
FRUIT: The fleshy fig has many small achenes inside.
OCCURRENCE: Native to the Mediterranean area; cultivated widely in yards, also commercially.
TOXICITY: The leaves contain psoralens which are furocoumarins (91).
SYMPTOMS: Itching, redness and skin blistering (90, 106, 142); photo-dermatitis (106, 126, 142, 171).
NOTES: The raw, unpeeled fruit sometimes irritates the lips, but when ripened it has very little of the milky sap in it. The fig was an important food and a sacred tree to many people from the Mediterranean to the Far East (113). The leaves, fruit and latex have also been used medicinally (142). While gathering figs, author James Preacher experienced redness and itching of the arms and neck when rubbed by the scabrous leaves.

Red Mulberry
Morus rubra L.
Moraceae

DESCRIPTION: Small or medium trees to 20 m; bark thin, with fissures and ridges.

LEAVES: Alternate, deciduous, 7.5–10 cm long, broadest near the base, entire, 1 or 3 lobed, toothed, thin and papery, with whitish hairs beneath.

FLOWERS: Small, greenish, in small round spikes; unisexual, on the same or different plants staminate with 4 sepals, 4 stamens; pistillate with 4 sepals and a 2-parted pistil; early Spring.

FRUIT: Cylindric, 2–3 cm long, fleshy, almost black when mature, sweet, edible; around 2 months after flowering.

OCCURRENCE: Native in rich woods, fence rows and waste areas of the eastern U.S.; white mulberry (*M. alba* L.) is introduced and naturalized in eastern North America; sometimes cultivated.

TOXICITY: The milky sap contains unidentified toxins.

SYMPTOMS: The ripe fruits are quite edible, but milky sap in immature fruits, stems and leaves causes stomach upset, hallucinations and nervous system stimulation (72, 142); the leaves may cause dermatitis (72, 142, 165); the pollen may cause hay fever (109).

NOTES: Legend has it that the red berries acquired their color after two young Babylonian lovers, Pyramus and Thisbe, bled and died under the tree (113). The Chinese began cultivating white mulberry as a food for silkworms about 2800 B.C.E (113), and early settlers imported it to this country in a failing attempt for the same reason (5). The bark of the root is a traditional treatment for ringworm in Europe (113). A drink from the fruit of white mulberry has been used for fever and as a laxative (102).

Stinging Nettle
Urtica dioica L.
Urticaceae

DESCRIPTION: Erect perennial herb; stems 4-angled, with stinging hairs; wood nettle (*Laportea*) is similar in appearance.

LEAVES: Opposite (*Laportea* has alternate), toothed.

FLOWERS: Small, greenish, in axillary panicles; staminate with 4 perianth parts, 4 stamens; pistillate with 4 perianth parts, ovary 1-celled; May-September.

FRUIT: Achenes.

OCCURRENCE: Waste areas, along road; North Carolina and north; cultivated.

TOXICITY: Stinging hairs on the leaves and stems contain irritant chemicals.

SYMPTOMS: Severe burning of the skin that persists usually less than 1 hour (72, 106, 142, 165).

NOTES: The stinging hairs have a bladder-like base filled with the irritant chemical, a slender capillary tube and a sharp tip that easily penetrates the skin. When the plant is touched, the hair bends, constricting the base and the chemical is forced into the skin through the capillary tube. Eaten raw, the plant causes gastrointestinal irritation but is an excellent potherb (142) high in vitamins A and C (5). The fresh juice of nettle has been used to promote milk flow in nursing mothers, for excessive menstrual flow, diarrhea and hemorrhoids (113). *Laportea* and *Cnidoscolus* (spurge nettle) have similar hairs as well.

Marijuana
Cannabis sativa L.
Cannabaceae

DESCRIPTION: Erect, coarse, annual herbs; stems rough, 1.8–3.6 m tall.
LEAVES: Palmately divided with 3–7 narrow, toothed leaflets; stalks long; alternate above, opposite below.
FLOWERS: Staminate many, in leafy, axillary panicles, sepals 5, petals none, stamens 5; pistillate in clusters on leafy branches.
FRUIT: Small achenes.
OCCURRENCE: Native to Asia; planted illegally throughout the U.S. for its narcotic effects.
TOXICITY: Various resins (mainly Δ^9 tetrahydrocannabinol or THC) occur throughout the plant (97).
SYMPTOMS: When smoked or otherwise ingested, the foliage causes euphoria, hallucinations, stupor and possible coma (72, 142); all in varying degrees depending upon the dosage; fresh leaves cause dermatitis in some people (72, 90, 106, 138, 142).
NOTES: This plant has been used for its narcotic properties for thousands of years. The Chinese used it as early as 2800 B.C.E. as a source of fiber. It spread to Europe around 500 C.E. and came to America with explorers (113).

Hops
Humulus lupulus L.
Cannabaceae

DESCRIPTION: Perennial, twining vine; stems rough, herbaceous, 3–10 m long.
LEAVES: Mostly opposite, palmately lobed and veined.
FLOWERS: Staminate in axillary panicles, sepals 5, petals none, stamens 5; pistillate in cone-like spikes; July-August.
FRUIT: Achenes; August-October.
OCCURRENCE: Native to America and Eurasia; naturalized in thickets, on marshy soil; cultivated on the Pacific coast as a source of hops for brewing.
TOXICITY: The plant has 0.3–1% volatile oils: mostly humulene, myrcene, β-caryophyllene and farnesene; 3–12% resinous bitter principles such as the α-bitter acids humulone, cohumulone, adhumulone, prehumulone and posthumulone; and β-bitter acids such as lupulone, colupulone and adlupulone (108).
SYMPTOMS: Some people get allergic contact dermatitis with eruptions upon contact with the leaves, flowers or pollen (106, 108, 165).
NOTES: Harvesters contracting dermatitis from this plant call it "hop pickers itch" (106, 109). The plant has been used in a poultice for skin boils and rheumatism (102). Internally, extracts have been used as a diuretic and to induce sleep. Excessive doses or extended use should be avoided (113).

Mistletoe
Phoradendron serotinum (Raf.) Johnston
Loranthaceae

DESCRIPTION: The state flower of Oklahoma; a parasitic, dioecious, evergreen shrub with green branches.
LEAVES: Opposite, simple, oblong, persistent, 2–13 cm long, to 4 cm wide.
FLOWERS: Small, regular; sepals 3–5, united at the base; petals none; stamens same number as sepals; October-November, sometimes throughout the Winter.
FRUIT: Small, 2–3 seeded, white berry; November-January, persisting into Spring.
OCCURRENCE: Native and parasitic on trees, usually hardwoods; New Jersey to Indiana and south.
TOXICITY: Berries and possibly foliage contain the toxic amines β-phenylethylamine and tyramine (49, 101, 102, 165).
SYMPTOMS: Gastrointestinal irritation, nausea, diarrhea, vomiting, slow pulse, labored breathing, delirium, hallucinations, sweating, pupil dilation, shock, cardiovascular collapse and death (72, 102, 142); dermatitis is also possible (106, 109, 142).
NOTES: In one case of poisoning, a 28-year-old female became ill after drinking mistletoe berry tea to bring on her menstrual period. Symptoms of vomiting, diarrhea and abdominal cramps continued for about 10 hours before medical help was sought. About 12 hours after ingesting the tea she was admitted to a hospital in deep shock, with dilated pupils, pale skin, profuse sweating, mental confusion, apathy, nausea and severe abdominal pains. Other symptoms were moderate dehydration, low temperature, unobtainable blood pressure and barely perceptible pulse. She died 10–15 minutes after arriving at the hospital, apparently of cardiovascular collapse. Gross examination revealed 130 ml of dark green liquid in her stomach (123).

Buckwheat

Fagopyrum sagittatum Gilib.
Polygonaceae

SYNONYMS: *F. esculentum* Moench.; *F. sagittatum* (L.) Karsten
DESCRIPTION: Erect, annual herb to 6 dm tall.
LEAVES: Simple, entire, broad and sagittate.
FLOWERS: Small, white in racemes; sepals petal-like; petals none; stamens 8; stigmas and styles 3; June-frost.
FRUIT: Three-sided nutlet.
OCCURRENCE: Native to Asia; cultivated as a grain for green manure; sometimes escaping along roadsides, into meadows and waste areas.
TOXICITY: A napthrodianthrone derivative known as fagopyrin causes photosensitization in animals (91).
SYMPTOMS: In humans: hay fever and possible skin irritation (171).
NOTES: Most human poisonings appear to come from an allergic response (101, 142). Susceptible persons get a rash from eating foods made from buckwheat flour (134). In one case, a 30-year-old bakery salesdeliveryman began noticing symptoms of nasal mucous inflammation when he was on his route; especially when clouds of dust were created by moving baskets in his truck or upon entering the bakery shop. Tests confirmed his allergy to buckwheat (24).

Rhubarb
Rheum rhaponticum L.
Polygonaceae

DESCRIPTION: Herbaceous perennial.
LEAVES: Large, to 4.6 dm long; heart-shaped, petiole red.
OCCURRENCE: Originating in Tibet and northwestern China; extensive domestication and selection have produced the variety now grown in vegetable gardens for the fleshy, red leaf stalks.
TOXICITY: The leaf blades contain dangerous levels of soluble oxalates and oxalic acid (101, 106, 109) which are very corrosive to the gastrointestinal tract (49).
SYMPTOMS: After about 24 hours: nausea, stomach cramps, vomiting, headache, somnolence, stupor, burning of the mouth and throat, electrolyte disturbances, muscle cramps, facial muscle twitching, kidney damage, acetone odor on the breath and death (49, 109); the leaves may cause skin irritation (90, 165).
NOTES: The leaf blades are poisonous even if cooked. During World War I, misinformed officials in Britain recommended that people use the blades to conserve food supplies. Several cases of poisoning occurred (6). In another case, a 4-year-old girl was given a considerable amount of the raw leaves by playmates in a garden. Early symptoms included nausea and lethargy. Her condition worsened the next morning and she began vomiting a dark green material. The child was given a sedative and taken to a hospital in a state of semi-stupor 24 hours after the symptoms began. Oxalic acid poisoning was confirmed by urinalysis which revealed oxaluria. In spite of treatment, she died within 2 days of the occurrence (164).

Dock, Sorrel
Rumex spp.
Polygonaceae

DESCRIPTION: Monoecious or dioecious perennial, biennial or winter annual herbs.

LEAVES: Alternate, oblong to lanceolate.

FLOWERS: Small and numerous in panicle-like racemes; sepals 6; stamens 6.

FRUIT: Achenes.

OCCURRENCE: Borders, fields and pastures; throughout the East; sometimes cultivated.

TOXICITY: The leaves contain varying amounts of soluble oxalates (109, 142).

SYMPTOMS: Refer to *Rheum* for details of oxalate poisoning; the crushed leaves and stems may cause allergic contact dermatitis (109, 165); the pollen may cause allergic respiratory reactions (109).

NOTES: Garden sorrel or sand begonia (*R. acetosa* L.) is a perennial species occasionally cultivated in the Northeast for greens. It contains oxalic acid and unless cooked causes delayed gastroenteric irritation (106). The young leaves of curled dock (*R. crispus* L.) are edible if the water is changed during cooking and a touch of baking soda is added to help neutralize the acid (142). Juice from this species has been used for ringworm, and a decoction of the fresh root has been used as a laxative (102). Mouth ulcers were treated with a tea from the flowers and leaves (113). If eaten by cows, the plant causes bad flavors in milk (129).

Mexican Tea, Wormseed

Chenopodium ambrosioides L.
Chenopodiaceae

DESCRIPTION: Erect annual or weakly perennial herb; stem branched, to 1.2 m tall.

LEAVES: Lanceolate, toothed, aromatic.

FLOWERS: Small, greenish, in a panicle; sepals 3–5, incurved over the fruit; petals none; stamens 5–6; ovary 1-celled; July-frost.

FRUIT: Bladdery utricle with 1 black seed.

OCCURRENCE: Waste areas throughout the U.S.; cultivated.

TOXICITY: The unsaturated terpene peroxide ascaridole (17–90%) is found in the oil of the seeds (101, 108, 142, 171). Others reported include: *l*-limonene, myrcene, *p*-cymene, α-terpinene and others (108).

SYMPTOMS: Overdoses of the oil have caused nausea, headache, hallucinations, upset stomach, bloody vomiting, weakness, convulsions, central nervous system disturbances, liver and kidney damage, paralysis, coma and death (142).

NOTES: Oil of chenopodium has been used to treat intestinal parasites, but safer, more efficient drugs are generally used today (171). The oil from wormwood [*C. amrosioides* L. var. *anthelmenticum* (L.) Gray] has been used for the same purpose in tropical America, but overdoses may result in death (109).

Four O'Clock
Mirabilis jalapa L.
Nyctaginaceae

DESCRIPTION: Showy, perennial herb with many varieties.

LEAVES: Opposite, smooth, pointed, 5–15 cm long.

FLOWERS: Trumpet-like, yellow, red, white or striped, opening in the afternoon; stamens 5; June-frost.

FRUIT: Small, dry, ribbed achene.

OCCURRENCE: Cultivated perennial in warm areas, an annual elsewhere; occasionally escaping.

TOXICITY: Roots and seeds contain unspecified toxins.

SYMPTOMS: Nausea, stomach pains, gastroenteritis, vomiting and diarrhea, especially in children (72, 90, 109, 128); handling the roots and seeds may also cause dermatitis (90).

Pokeberry, Poke Salet

Phytolacca americana L.
Phytolaccaceae

DESCRIPTION: Erect, stout, perennial herbs; stems turning purple to red when mature; to 2.4 m tall.

LEAVES: Alternate, glabrous, simple, entire, to 30 cm long.

FLOWERS: Small, white, in racemes; sepals 5, persisting in fruit; petals none; stamens 5–30; May-frost.

FRUIT: Purple berry; seeds lustrous black.

OCCURRENCE: Native along fences, roadsides, in fields, waste areas of the eastern U.S. and Canada; cultivated.

TOXICITY: A saponin phytolaccine, phytolaccotoxin, phytolaccic acid and phytolaccigenin occur mostly in the foliage and roots (109, 128, 142).

SYMPTOMS: Burning sensation of the mouth and throat, salivation, perspiration, gastrointestinal cramps, nausea, vomiting, labored breathing, weakness, weak pulse, visual disturbances, spasms, convulsions, respiratory failure and death (72, 101, 109, 142).

NOTES: Indians used small doses of the crushed and roasted root as a blood purifier (159), and the powdered root in a poultice to treat cancer (102). The dried root is still used in Appalachia in treating hemorrhoids (5). The Pamunkey Indians of Virginia used the boiled leaves for rheumatism (109). This is a very dangerous native plant because it is often eaten without proper cooking or the roots are eaten like potatoes. When the young leaves and stems are eaten as greens, the water should be changed at least twice during cooking. The raw berries are toxic, but safe when cooked (142). In one case of poisoning, a man misidentified poke root as horseradish. Though he swallowed only a small amount, it was very bitter and burned his mouth. About 1.5 hours later, he began yawning and coughing and had stomach cramps and vomiting, the last two recurring every 5–10 minutes. Two hours after ingestion, he had visual disturbances, could not stand or even sit, had excessive salivation, convulsive tremors, prickly sensations, coldness with clammy perspiration, a sense of suffocating, a dull aching in the loins and contracted pupils. He received relief after a few minutes with morphine sulfate and atropine, but vomiting recurred hourly for the next 11 hours. Diarrhea and a bitter taste persisted for about 2 days before recovery (57). Mature plants should be handled with gloves for toxins in the juice may enter the body through cuts and abrasions (109).

Corn Cockle
Agrostemma githago L.
Caryophyllaceae

DESCRIPTION: Erect, winter annual; taproot shallow; stems rough, to 9 dm tall.

LEAVES: Opposite, hairy, 8–12 cm long, jointed at the base, linear or lanceolate.

FLOWERS: Solitary or in cymes, purplish-red; sepals united, with lobes longer than the tube; stamens 10; stigmas and styles 5; May-September.

FRUIT: Capsule with one locule; seeds black.

OCCURRENCE: Native to Eurasia; often present with fall grown grain crops such as winter wheat and rye; some species are cultivated as ornamentals.

TOXICITY: Seeds contain the saponin githagin (72, 142) which is 5–7% of their weight (101) and the sapogenin githagenin (109, 142).

SYMPTOMS: Severe gastroenteritis, nausea, abdominal pain, dizziness, weakness, vomiting, diarrhea and slow, labored breathing (72, 109, 142).

NOTES: Modern screening machines are now used to remove contaminant seeds such as corn cockle (101), but home grown grains are dangerous if contaminated (72, 109). Wheat screenings ground up at grain elevators for use as animal feed sometimes contain this species. Dust from the grinding process caused one man problems for several years. After the first year, he had hay fever sometimes followed by asthma attacks. The patient was de-sensitized over a period of two years by very dilute concentrations of corn cockle extract (64).

Bouncing Bet
Saponaria officinalis L.
Caryophyllaceae

DESCRIPTION: Erect, perennial herbs; stems jointed.

LEAVES: Opposite, simple, sessile, entire, lanceolate, palmately veined.

FLOWERS: In clusters; sepals 5, tubular; petals 5, white, pink or red; stamens 10; styles usually 2; May-October.

FRUIT: Capsule; seeds with short, wart-like knobs.

OCCURRENCE: Disturbed, waste areas, roadsides, meadows; throughout the U.S.; sometimes cultivated.

TOXICITY: The seeds contain githagenin (109); saponins known as saporubin and saporubinic acid are reported in the roots (171).

SYMPTOMS: Gastroenteric irritation, chiefly of the intestinal mucosa (106) and destruction of red blood cells (93).

NOTES: Many folk names associated with the plant such as soapwort and soaproot refer to the plant's ability to form a soap-like lather in water caused by saponins. Early Anglo-Saxons used it for liver ailments, earache, swelling and leprosy (159). Decoctions containing the root have been used to remove discoloration around black or bruised eyes (109). Most problems arise from consumption of contaminated home-grown grains because commercial shipments are screened to remove such seeds (106).

Monkshood, Wolfbane

Aconitum spp.
Ranunculaceae

DESCRIPTION: Showy, perennial herbs; stems ascending or trailing, to 1.8 m tall; roots tuberous or thick and fibrous.

LEAVES: Alternate, palmate, 3–9 lobed.

FLOWERS: In panicles or terminal racemes; sepals 5, irregular, petal-like; petals 2–5, the upper large and hooded, white, blue or yellow; stamens many; pistils 3–5.

FRUIT: Follicles with several seeds.

OCCURRENCE: Native and introduced; rich woods in mountains and Piedmont; the European species *A. napellus* L. is cultivated in gardens of the U.S. and Canada.

TOXICITY: All parts of all species contain the alkaloid aconitine (72, 106, 109, 128). The root is the most dangerous part and most toxic before flowering (109, 134). Other toxins reported include picratine, picraconitine, aconiine and napelline (108). The lethal dose of aconitine is 2–4 mg and 15–30 ml of the tincture (10).

SYMPTOMS: Nausea, anxiety, vomiting, diarrhea, weakness, prostration, restlessness, salivation, numbness, vertigo, prickling of the skin, impaired speech and vision, chest pain, spasms, weak pulse, convulsions, paralysis of lower then upper extremities, respiratory paralysis, coma and death a few hours after consuming the leaves, flowers or roots (72, 101, 109, 142); if rubbed on the skin, the sap causes numbness and tingling (134).

NOTES: People of the Isle of Ceos used it to dispose of their elderly when they were no longer useful (103, 134). The Greeks and Romans used it to poison the water supply of enemies (113) and to make poison spears and arrows (116). The Roman Emperor Trajan (78–117 C.E.) prohibited the cultivation of aconite on penalty of death (93).

Baneberry
Actea spp.
Ranunculaceae

DESCRIPTION: Perennial herbs, 1 m or so tall; rootstock thick.
LEAVES: Two–3 ternately compound; leaflets ovate, toothed.
FLOWERS: In a short, thick terminal raceme; sepals 4–5, falling early; petals 4–10, small, clawed, red or white; stamens many; pistil 1.
FRUIT: Berry-like, indehiscent fruit with several seeds.
OCCURRENCE: White baneberry (*A. pachypoda* Elliott) and red baneberry [*A. rubra* (Ait.) Willd.] are native in rich woods of the East; cultivated.
TOXICITY: Both species are toxic (72, 138). The roots are purgative, irritant and emetic (101). All parts contain the innocuous glycoside ranunculin, that readily breaks down to form its aglycone protoanemonin, a volatile irritant oil (101, 106, 142).
SYMPTOMS: Dizziness, headache, increased pulse, vomiting, diarrhea, gastroenteritis; rarely convulsions and death (101, 109, 142); dermatitis.
NOTES: The question of baneberry toxicity was settled in 1903, by Alice E. Bacon, of Bradford, Vermont, in a curious experiment with the fruit. Failing to find adequate information on the toxicity of the plant, she took a small amount of the berries one day after the noon meal. The only effect was a slight burning of the stomach. Two days later, allowing the first dose to be eliminated, she took twice the original amount. In about a half hour, the pulse quickened and there was severe burning of the stomach. The symptoms lasted about 15 minutes. Two days later twice the second amount (probably 6) was taken. In a short time, she experienced intense hallucinogenic displays with various blue shapes followed by confusion, incoherency and dizziness. Other symptoms she reported were parched throat, difficult swallowing, an intense burning in the stomach with gaseous belches, followed by abdominal colic pains, a pulse of 125 and an unpleasant heart fluttering. The symptoms lasted about an hour, with complete recovery in about 3 hours (13).

Windflower

Anemone spp.
Ranunculaceae

DESCRIPTION: Perennial herbs with rhizomes or tubers.
LEAVES: Stem leaves 2–9 together, whorled or opposite, some dissected, appearing compound.
FLOWERS: Solitary or umbellate; sepals few-many, petal-like white-greenish; petals none or like abortive stamens; stamens many.
FRUIT: Achenes.
OCCURRENCE: Rich woods, Mountains and Piedmont: some in coastal areas; some cultivated.
TOXICITY: The innocuous glycoside ranunculin breaks down by enzymatic action caused by bruising or injury to form the irritant aglycone protoanemonin (101, 106, 109, 128).
SYMPTOMS: Irritation of mucous membranes, burning of the throat, vomiting, bloody diarrhea, dizziness, fainting and convulsions; further symptoms may be watery skin blisters, excessive urination with blood (142); it may also cause skin, eye and respiratory irritation (106, 109, 142). These symptoms apply to all plants with protoanemonin including *Actea, Anemone, Caltha, Clematis* and *Ranunculus.*
NOTES: Early superstition held that even wind blowing over a field of the flowers became poisonous (48). The toxicity of protoanemonin is removed by cooking or drying (49, 106, 142).

Marsh Marigold, Cowslip

Caltha palustris L.
Ranunculaceae

DESCRIPTION: Glabrous, perennial herb; stem hollow, 3–6 dm tall; rootstock short, stout.
LEAVES: Heart-shaped, mostly basal, 18 cm wide, 5–15 cm long; dentate or entire.
FLOWERS: Yellow, usually 2 per bract; sepals 5–6, conspicuously veined; petals none; stamens many; April-June.
FRUIT: Follicles with red, obovoid seeds.
OCCURRENCE: Swamps, woods, marshy meadows; from Canada to the Carolinas and westward; sometimes cultivated.
TOXICITY: The innocuous glycoside ranunculin breaks down by enzymatic action caused by injury or bruising of plant tissues to form the aglycone protoanemonin, an irritant volatile oil (93, 101, 106, 109, 134).
SYMPTOMS: See *Anemone* for details.
NOTES: Cowslips are sometimes used as greens in early spring after thorough cooking to drive off the toxin (106). One common name *verrucaria* refers to the plant's use in removing warts (5, 93). A tincture from the plant has also been used for anemia (134). It causes a bad flavor in milk if eaten by cows (129). In Anglo-Saxon, the word for cowslip was *cuslyppe; cu* for cow and *slyppe* for slop. The literal translation is "cow dung" (48).

Leather Flower, Virgin's Bower

Clematis spp.
Ranunculaceae

DESCRIPTION: Erect or ascending, slightly woody, vines.
LEAVES: Opposite, simple or pinnate.
FLOWERS: Sepals 4–5, white, purple or blue; petals none; stamens many; plumose styles persistent in fruit.
FRUIT: Achenes.
OCCURRENCE: Native and introduced; woods and borders; some cultivated.
TOXICITY: The glycoside ranunculin breaks down to form the irritant aglycone protoanemonin (106, 109).
SYMPTOMS: See *Anemone* for details.
NOTES: The leaves and flowers of some species have been used to promote perspiration, as a diuretic, for headache and as a stimulant. Externally, it has been used for skin disorders (113). In one case of dermatitis, a woman experienced vesicular eruptions of the face, neck, chest, hands and arms; the eyelids were swollen and accompanied by conjunctivitis. These symptoms had been recurring for 7 summers, beginning in May and lasting until November. Skin patch tests were positive only for *Clematis*. When the plant was removed from her porch she was well in 3 weeks and remained so (107).

Larkspur
Delphinium spp.
Ranunculaceae

DESCRIPTION: Erect, annual or perennial herbs; rhizomatous, single or clustered tubers.

LEAVES: Alternate or clustered near the ground, palmate to deeply divided with 3–5 divisions.

FLOWERS: Showy, in erect terminal racemes or panicles; sepals 5, red, white, purple or blue, the upper spurred; petals smaller, usually 4, the upper pair projecting back into the sepal spurs.

FRUIT: Follicles with many seeds.

OCCURRENCE: Native and introduced; in woods and waste areas; some cultivated.

TOXICITY: The alkaloids delphinine, delphinoidine, ajacine, delphisine and others are found in the seeds and young plants (49, 72, 109, 142).

SYMPTOMS: Burning of the mouth, tingling skin, nausea, stomach upset, abdominal pain, weak pulse, labored respiration, nervousness, depression or excitement and death (72, 109, 142); dermatitis (72, 90, 142, 165).

NOTES: These plants have been considered medicinal and poisonous for centuries. An old pharmaceutical preparation from the seed called "larkspur lotion" was once used to treat head lice, but is now considered too dangerous and is no longer available commercially (49). Toxicity of the plants decreases with age (72).

Christmas Rose
Helleborus niger L.
Ranunculaceae

DESCRIPTION: Erect, evergreen, perennial herb; rootstock short, black.
LEAVES: Compound with 7–9 leaflets.
FLOWERS: Usually solitary on red-spotted peduncles, white sometimes with pink; sepals 5, petaloid; petals small, tubular, clawed; appearing in Winter if protected from extreme cold.
FRUIT: Dry, single-celled, splitting along one seam.
OCCURRENCE: Native to Europe; introduced and cultivated throughout the U.S.
TOXICITY: Roots and leaves contain the cardiac glycoside hellebrin and others (109).
SYMPTOMS: Slow, irregular pulse, salivation, weakness, abdominal pain, vomiting, purging, labored breathing, convulsions, respiratory failure and sometimes death (4, 134); dermatitis in some people (129).
NOTES: Legend has it that a country girl named Madelon, who visited the Christ child in Bethlehem with the shepherds, was sad because she had no gift for him. An angel, seeing her plight, led her outside and touched the ground causing the first Christmas rose to appear (113). Hellebore was one of the four classic poisons with hemlock, nightshade and aconite. Its use in treating intestinal worms lasted into the 18th century. A warning in one advertisement said in effect that if it did not kill the patient, it would certainly kill the worms, but sometimes both. It has also been used as an emetic, diuretic, emmenagoge, irritant and cardiac (113). The ancients used hellebore externally for lice (93) and it has been used in Africa as an arrow poison (171). Drying or storage does not destroy the toxins (134).

Golden Seal
Hydrastis canadensis L.
Ranunculaceae

DESCRIPTION: Erect, perennial herb, rhizones yellow.

LEAVES: Hairy, cordate, 3–7 lobed; near the top of the unbranched stem; serrate.

FLOWERS: Solitary; sepals 3, petal-like, falling early; petals none; stamens and carpels many; April.

FRUIT: Dark red berry; May-June.

OCCURRENCE: Native in rich woods; North Carolina and north; may be cultivated.

TOXICITY: Isoquinoline alkaloids such as hydrastine (1.5–4%) and berberine (0.5–6%) with lesser amounts of canadine, candaline and others (108).

SYMPTOMS: Irritation of mucous membranes with ulcerous secretions (113).

NOTES: Indians rubbed the yellow sap on their bodies with bear grease as an insect repellent (102). Infusions of the root were used by Indians of eastern America for liver and stomach ailments. Midwestern Indians and early settlers steeped the dried root in water and applied this liquid to inflamed eyes (109). The plant has also been used as a laxative (113).

Buttercup, Crowfoot

Ranunculus spp.
Ranunculaceae

DESCRIPTION: Perennial or annual herbs.
LEAVES: Stem leaves alternate, entire, lobed or divided, palmately veined; sometimes with numerous basal leaves.
FLOWERS: Solitary, terminal or in corymbose clusters; sepals usually 5; petals often 5, yellow; stamens many.
FRUIT: Small achenes.
OCCURRENCE: Native and introduced; in various habitats; some cultivated.
TOXICITY: All parts contain the innocuous glycoside ranunculin which yields the irritant aglycone protoanemonin by enzymatic action when the plant is crushed or otherwise damaged (49, 101, 142); the highest concentration occurs during flowering (49).
SYMPTOMS: See *Anemone* for details of poisoning; the pollen may cause respiratory irritation (109).
NOTES: Beggars once used crowfoot to create sores on their limbs to arouse sympathy (134). Cows eating the plant produce bitter tasting milk that may have a reddish tinge (129).

Pawpaw
Asimina triloba (L.) Dunal
Annonaceae

DESCRIPTION: Small understory tree or large shrub; to 10 m tall; branches pubescent, Winter buds hairy.

LEAVES: Alternate, deciduous, large, simple, glabrous above, hairy beneath.

FLOWERS: Solitary along stems produced in the previous season; stalked, bell-shaped; sepals 3; petals 6, purple; stamens many; pistils 3–15; Spring.

FRUIT: Irregularly cylindrical, aromatic; black, soft and fleshy at maturity; seeds large, flattened; late Summer, early Fall.

OCCURRENCE: Rich, moist soils such as found in river bottoms and valleys, sometimes dense thickets; New York to Nebraska and south; sometimes cultivated.

TOXICITY: Undefined.

SYMPTOMS: Severe gastrointestinal problems after eating the fruit in some people (90, 142); occasional dermatitis from handling it (72, 90, 106, 142, 165).

NOTES: The fruit juice has been used for intestinal worms; the seeds to induce vomiting; powdered seeds have been used for head lice (5, 102).

Blue Cohosh, Papoose Root

Caulophyllum thalictroides (L.) Michx.
Berberidaceae

DESCRIPTION: Glabrous, rhizomatous, perennial herb; stem simple, erect, to 6 dm tall.

LEAVES: Large, divided, unstalked.

FLOWERS: In racemes or panicles; each with 6 petal-like sepals; 6 reduced petals; stamens 6; April-early June.

FRUIT: With 1–2 drupe-like, blue, naked seeds.

OCCURRENCE: Native in rich woods from Canada south to Missouri; in Mountains and Piedmont to South Carolina and Alabama; also cultivated.

TOXICITY: Leaves and seeds contain the alkaloid methyl-cytisine and saponins (72, 106, 109); rhizomes and roots contain these and others including baptifoline, anagyrine and magnaflorine (108).

SYMPTOMS: Gastroenteric irritant, chiefly of the intestinal mucosa (106); berries, leaves and roots may cause allergic contact dermatitis (72, 129, 138, 165).

NOTES: The roasted seeds have been used as a coffee substitute, but raw seeds have poisoned children (72). Indians used the powdered root to induce menstruation and labor (102).

Mayapple
Podophyllum peltatum L.
Berberidaceae

DESCRIPTION: Glabrous, perennial herb with simple stems; rootstock stout.
LEAVES: Two that are large, rounded, palmately divided with 5–9 parts.
FLOWERS: Solitary, nodding, white; sepals 6; petals 6–9; stamens 12–18; stigma sessile; April-early June.
FRUIT: Large, fleshy, red or yellow berry.
OCCURRENCE: Open fields, pastures, woods; throughout North America, often forming dense colonies; sometimes cultivated.
TOXICITY: The resin podophyllin in all parts, but mostly the rhizomes, contains lignans such as podophyllotoxin, an anti-mitosis agent (97, 109) and others. If eaten by the mother during pregnancy, the effects on mitosis could cause genetic defects in the human fetus (72).
SYMPTOMS: Small amounts of the unripe fruit cause gastroenteritis, diarrhea, vomiting, and abdominal pain (101, 109, 142). Larger amounts will cause dizziness, headache, fever, labored breathing, rapid pulse, low blood pressure, coma and death (142); the roots may cause dermatitis (72, 129, 142, 165).
NOTES: Poisoning does not often occur for the ripe fruit is safe in small amounts. It has even been made into jelly (129). Children, however have been poisoned by eating too many green fruits (109). Toxicity is usually seen in misuse of medical preparations for constipation (49, 109). The roots have been used for fever, cough, liver problems, jaundice and syphilis (102). Workers handling the powdered rhizome commercially sometimes experience conjunctivitis keratitis and ulcerative skin lesions (101). The resin has also been used to treat warts. The following case illustrates the effects when given internally. On January 19, 1964,

a 19-year-old female in the 26th week of pregnancy was taken to a hospital with complaints of frequent urination with burning pains. The admission diagnosis was Condylomata and Cystititis. Twenty percent podophyllum resin in compound benzoin was applied locally to treat the first condition. At 6 P.M., she was given a teaspoon of the resin orally by mistake. About 4 hours later, she began coughing and vomiting. Treatment ensued for 3 days, after which time she was much improved. In another case, death followed the use of a local application of 25% resin ointment for a condyloma acuminatum of the vulva. Vomiting and respiratory stimulation occurred, followed by coma. Death occurred on the 7th day after application of the resin (17).

Moonseed

Menispermum canadense L.
Menispermaceae

DESCRIPTION: Glabrous or pubescent, woody, twining vine.

LEAVES: Alternate, ovate, palmately lobed with 3–5 segments; glabrous above, glaucous beneath.

FLOWERS: In panicles; sepals and petals very small; stamens 12–24; June-August.

FRUIT: Black, globular drupe; surface with whitish wax; in grape-like clusters; each fruit with 1 grooved, crescent-shaped seed; the fruits have been confused with grapes which have several seeds, rather than one.

OCCURRENCE: Native in low woods, along stream banks, fence rows; Canada south to Georgia and Oklahoma; cultivated in the north for its foliage.

TOXICITY: All parts are toxic (106, 138); roots and fruits have the isoquinoline alkaloid dauricine that has curare-like action (109).

SYMPTOMS: Pictrotoxin-like convulsions and death (106); the spiny fruit pits may cause intestinal irritation (129).

NOTES: Children have been killed eating the fruits. Birds eat them without harm, but they readily consume many fruits and seeds poisonous to other animals (72). The fresh roots yield a tincture that has been used as a laxative, for venereal disease and gout (102).

Mexican Prickle Poppy, Thorn Apple

Argemone mexicana L.
Papaveraceae

DESCRIPTION: Showy, erect, prickly, annual herb; sap yellow; stems branched, to 9 dm tall.

LEAVES: Alternate, sessile, base, clasping the stem, lanceolate, dentate.

FLOWERS: Almost sessile, solitary; sepals 2–3, hooded; petals 4–6; stamens many; carpels 4–6; April-September.

FRUIT: Prickly capsule, with 3–6 valves at the top.

OCCURRENCE: In waste areas, along sunny roadsides; Florida to Arizona and north to Pennsylvania; sometimes cultivated.

TOXICITY: Alkaloids berberine, protopine, sanguinarine and dihydrosanguinarine are found in the leaves and seeds (72, 109, 142).

SYMPTOMS: Vomiting, diarrhea, visual disturbances, abdominal swelling, fainting, coma; circulatory, kidney and nervous disorders may occur (72, 142); the prickles may cause skin irritation (142, 165).

NOTES: Problems sometimes arise when the seeds contaminate home-grown grains (72, 109). Contaminated wheat caused epidemic dropsy in India (101, 109). The toxins may be transmitted in the milk of dairy animals not yet showing signs of poisoning (101). The seed oil is widely used in parts of India and is of concern because it has high levels of sanguinarine, a carcinogenic benzophenanthridine alkaloid (100). This toxin has also been found to be transmitted from the diet into the milk of goats and rabbits (69).

Celandine, Rock Poppy
Chelidonium majus L.
Papaveraceae

DESCRIPTION: Erect or decumbent biennial; sap orange to yellow; a monotypic genus.
LEAVES: Pinnately divided; several on the flowering stem; glaucous beneath.
FLOWERS: Regular, in umbels: sepals 2; petals 4, yellow; stamens many; pistil 1; March-August.
FRUIT: Two-valved, many-seeded capsule; opening from the bottom up.
OCCURRENCE: Native to Europe and Asia; cultivated and naturalized in moist soil from Canada to North Carolina and Missouri; sometimes cultivated.
TOXICITY: Underground portions of the plant contain chelidonine, sanguinarine, chelerythrine, protopine, berberine, tetrahydrocoptisine and others (101, 109, 129); seeds, leaves and stems contain berberine, protopine, sanguinarine and dihydrosanguinarine (18, 72).
SYMPTOMS: Nausea, vomiting, bloody diarrhea, abdominal swelling, giddiness, painful urination, circulatory disorders, numbness, fainting, coma and death (4, 72); the sap may cause skin irritation with skin blisters (4).
NOTES: The sap has been used externally in Europe to treat warts (109, 134). Used internally, it is toxic in overdoses (106) and poisoning has been reported in Europe (101).

Poppy, Opium Poppy
Papaver somniferum L.
Papaveraceae

DESCRIPTION: Erect annual; stem simple or branched; to 1 m tall; with a taproot.

LEAVES: Ovate to oblong, lobed, margins wavy; glaucous; upper leaves clasping the stem.

FLOWERS: Large; sepals 2–3; petals white or pale lilac, 9 or more; stamens many; stigma sessile.

FRUIT: Capsule: seeds black or white, small.

OCCURRENCE: Native to Europe and Asia; now controlled by federal law, only licensed persons may cultivate the plant; sometimes persisting in old gardens.

TOXICITY: The dried latex known as opium has isoquinoline alkaloids such as morphine, codeine, and papaverine. Others are laudanine, thebaine, laudonosine, chelidonine, sanguinarine, berberine, protopine and others (39, 97, 109, 129, 142, 163).

SYMPTOMS: The unripe fruit can cause stupor, shallow breathing, euphoria, central nervous system, respiratory and circulatory depression, coma and death (72, 142).

NOTES: The plant has been used as an opiate for thousands of years. The Arab physician and philosopher Avicenna died of opium intoxication in 1037 C.E. (22). Deaths are still recorded from addiction and misuse of opium. The seeds used as a topping for bread have only a trace of the alkaloids and are safe (72, 142).

Bloodroot
Sanguinaria canadensis L.
Papaveraceae

DESCRIPTION: Perennial herb with reddish sap; with a fleshy, red underground stem; a monotypic genus.

LEAVES: Solitary, palmately lobed, kidney-shaped, glabrous.

FLOWERS: Solitary on a scape; sepals 2, falling early; petals white, 8–12, conspicuous before the leaf unrolls; stamens usually 2× as many as petals; March-May.

FRUIT: Many-seeded, 2-celled capsule.

OCCURRENCE: Native in rich woods, fence-rows; southern Canada to Florida and Texas; sometimes transplanted into wildflower gardens.

TOXICITY: The isoquinoline alkaloid sanguinarine (about 1%) and numerous others, including sanguidimerine, chelerythrine and protopine occur throughout the plant, especially in the rhizomes (108).

SYMPTOMS: Overdoses of medical preparations have caused burning of mucous membranes, nausea, vomiting, labored breathing, dilated pupils, fainting, shock and coma, muscular and cardiac failure and death (72, 142); the juice is a skin and eye irritant (72, 90, 109, 142, 165).

NOTES: Indians used the root to induce vomiting, to cure sore throat, for fevers and rheumatism (102).

Dutchman's Breeches

Dicentra spp.
Fumariaceae

DESCRIPTION: Delicate, scapose, perennial herbs from rhizomatous bulbet-bearing rootstocks.
LEAVES: Basal, broadly triangular, ternately compound, lace-like.
FLOWERS: Nodding, in racemes or panicles; sepals 2, reduced, sometimes falling early; petals 4, the outer spurred, dark pink, flesh-colored to purple, white or greenish-white.
FRUIT: Valvate capsules, 10–20 seeded; seeds black.
OCCURRENCE: Rich moist woods, shaded ledges; some cultivated.
TOXICITY: Isoquinoline alkaloids such as protopine and others occur in all parts (72, 163).
SYMPTOMS: Labored breathing, trembling, incoordination, convulsions; large amounts may be fatal; dermatitis (72).
NOTES: At one time, the underground parts of *D. cucullaria* (L.) Bernh. were used as a tonic, to increase urine flow and treat venereal diseases (102).

Hydrangea
Hydrangea spp.
Saxifragaceae

Hydrangea (domestic)

Hydrangea quercifolia

DESCRIPTION: Shrubs up to 3 m tall.

LEAVES: Opposite, simple, exstipulate, glabrous or pubescent beneath.

FLOWERS: In compound cymes; calyx tubular, 4–5 lobed, 8–10 ribbed; petals 5, white; stamens 8–10, sometimes exserted; marginal flowers of the cluster usually sterile with a showy 3–4 lobed calyx.

FRUIT: Capsules, strongly ribbed.

OCCURRENCE: Native and introduced: steep slopes, New York to Illinois and south; some cultivated.

TOXICITY: Leaves, branches and buds contain hydrangin, a cyanogenic glycoside (72, 109, 142) and possibly other toxins.

SYMPTOMS: Nausea, vomiting, diarrhea, gastroenteric distress, labored breathing, weakness, stupor, coma, convulsions and fibrillary twitching (49, 109, 142).

NOTES: A Florida family experienced nausea and gastroenteritis when they added garden hydrangea (*H. macrophylla* Seringe) buds to a salad (101, 142). The roots of smooth hydrangea (*H. quercifolia* Bart.) were used by pioneers to treat indigestion (109). The roots of various species were also used for kidney stones, as a diuretic, tonic and laxative (102). If eaten by cows, the plant will cause bad flavors in milk (129).

Apple
Malus sylvestris Mill.
Rosaceae

DESCRIPTION: Small tree or shrub; glabrous or pubescent when young.
LEAVES: Simple, ovate to elliptic, crenate-serrate, with small adnate stipules.
FLOWERS: Four-7, in umbellate clusters; pedicels and calyx tube exterior glabrous; petals white, with some pink; stamens many; styles 5.
FRUIT: Fleshy pome with several seeds.
OCCURRENCE: Native to Eurasia; introduced and cultivated for the fruit.
TOXICITY: The fruit pulp is harmless, but the seeds contain the glycoside amygdalin, which yields HCN and benzaldehyde upon hydrolysis (90, 101, 109, 142) which could be brought on by being crushed or eaten.
SYMPTOMS: Refer to *Prunus* for details of cyanide poisoning.
NOTES: Seeds cause poisoning only when consumed in large amounts. HCN prevents cells from using oxygen even though it is abundant in the tissues. Amygdalin is also found in the seeds of pears (*Pyrus spp.*) (109).

Cherry

Prunus spp.
Rosaceae

DESCRIPTION: Shrubs or trees.
LEAVES: Alternate, simple, usually deciduous, generally toothed, sometimes with glands on leaf stalks.
FLOWERS: Solitary or in few-flowered clusters, white or pink, perfect; calyx 5-lobed, bell-tubular shaped; petals 5; stamens 15–30; pistil 1.
FRUIT: Drupes, round to oval-shaped, pulp dry or juicy, surrounding a hard pit; the seed is inside the pit.
OCCURRENCE: Native and introduced; some cultivated.
TOXICITY: Seeds contain the cyanogenic glycoside amygdalin and leaves contain prunaisin which yield HCN upon hydrolysis (145). The unbroken seeds are harmless (106) and toxicity in leaves is brought on by wilting. The lethal dose is about 50 mg in adults and 20 mg in children (72).
SYMPTOMS: Small amounts cause gasping, weak and irregular respiration, weakness, excitement, depression, staggering, pupil dilation, twitching, spasms, convulsions, coma, acetone odor on the breath, respiratory failure and death (10, 142); lethal amounts may cause spasms, respiratory failure and death within 1 hour (142).
NOTES: Children have been poisoned eating the seeds, drinking tea from the leaves and chewing the twigs of black cherry (*P. serotina* Ehr.) (109). A tea from the bark was used by Indians and in Appalachia to treat coughs and colds (102, 109). Some parts of related species should also be considered toxic including cherry laurel, plum, bitter almond, peach and apricot (109). The controversial, reputedly cancer-curing drug *laetrile* is taken from the seeds of apricots, peaches, and related fruits

(108). Peach leaves have been used for vomiting and morning sickness during pregnancy, but the dosage must be regulated to avoid complications (113). The seeds of bitter almond [*P. amygdalus* Batsch. var. *amara* (DC) Focke] contain amygdalin and prunaisin and should not be eaten. Sweet almonds [*P. amygdalus* Batsch. var. *dulcis* (DC) Koehne] do not contain the toxins and are safe as food (108). In June, 1978, a 49-year-old woman ate 20–40 apricot pits she purchased at a health food store. Within ½ hour, she had headache, weakness, disorientation and nausea. She vomited once or twice expelling a large part of the kernels. Later analysis of the seeds showed a cyanide content of 409 mg/100 g of moist seeds, which far exceeds the minimum adult lethal dose for HCN. Vomiting most of the seeds, reduced the available dose and saved her life, together with antidotal and supportive care (151).

Rosary Pea, Precatory Bean

Abrus precatorius L.

Leguminoseae

DESCRIPTION: Perennial, climbing, twining herb; stem gray to green.

LEAVES: Pinnately divided; 8–15 pairs of leaflets.

FLOWERS: Small, purplish-red, in axillary racemes.

FRUIT: Hairy, many-seeded pod, 4 cm long; seeds glossy, red and black.

OCCURRENCE: Native to southeast Asia; cultivated in the southern U.S. as an ornamental; a weed of fence rows and cultivated fields.

TOXICITY: The beans contain abric acid and abrin, a toxalbumin (72, 101, 106, 128, 142).

SYMPTOMS: Nausea, vomiting, stomach pain, colic, bloody diarrhea, anuria, cramps, pupil dilation, fever, thirst, burning in the throat, headache, shock and death (139).

NOTES: Swallowing the bean whole is unlikely to cause poisoning for the hard seedcoat is unaffected by digestive juices (128, 171). Soft, immature seeds though, are dangerous, especially to children (128). One to a few beans may be lethal if chewed and swallowed (139). The scarlet bean is sometimes used in handicrafts such as beads and other jewelry. Thus, people unaware of the danger often bring them into areas where poisonings would not normally occur. Four such cases occured in Boston in 1963, where the beans were used in necklaces. A fifth case involved their use in a Christmas tree ornament (73). Abrin is a large protein molecule. Like other proteins, it is denatured by extensive heating and thus rendered nontoxic (165).

Coffee Senna
Cassia occidentalis L.
Leguminoseae

DESCRIPTION: Annual herb; to 2 m tall.
LEAVES: Pinnate, glandular; leaflets 8–12, ovate to lanceolate, acute-acuminate, 2–9 cm long; stipules falling early.
FLOWERS: Axillary, solitary or in few-flowered racemes; petals few; fertile stamens 6–7, of 2 sizes, sterile stames 3–4; July-August.
FRUIT: Legume, straight to slightly curved; to 14 cm long.
OCCURRENCE: In waste areas throughout the East; also cultivated.
TOXICITY: The seeds contain embodin glycosides (anthraquinones) and the toxalbumin chrysarobin; roots, leaves and fruits contain oxymethyl-anthraquinone (28, 109, 128, 142).
NOTES: The seeds have been used to treat ringworm and eczema in the tropics (109). Roasted seeds have been used as a coffee substitute (28, 171). An extract from the leaves has been used internally as a laxative, but large amounts are toxic (142). Overdoses of other herbal medicines from various species of the genus have caused human deaths (109).

Rattlebox

Crotolaria spp.
Leguminoseae

DESCRIPTION: Annual or perennial herbs.
LEAVES: Simple or 3–7 palmately divided, some with arrow-like stipules.
FLOWERS: Pea-like, yellow, in terminal racemes; calyx bell-shaped; standard large, heart-shaped, keel scythe-shaped; stamens 10.
FRUIT: Inflated, leathery pods, with loose seeds rattling inside when dry.
OCCURRENCE: Native and introduced; open woods, margins, sandy soils, roadsides, fields; in most of the East; some cultivated.
TOXICITY: The pyrrolizidine alkaloid monocrotaline and others (97, 109, 163); monocrotaline, spectabiline, retrorsine and others have been isolated from *C. spectabilis* Roth (109, 128, 142).
SYMPTOMS: Low level ingestions of pyrrolizidine alkaloid containing teas may cause lung ailments; large amounts cause liver damage (3).
NOTES: Some species have caused human poisoning by contaminating flour in other countries (101). Herbal extracts from the seeds and foliage have caused liver damage among children in Jamaica. It is termed *veno-occlusive disease* because of the serious damage to small branches of the hepatic venous system (3). In 1976, an epidemic of this condition was seen in Afghanistan when about 1600 people became ill after eating bread made from wheat that was contaminated with *Crotolaria* and *Heliotropium*. Many of them died. A similar outbreak occurred that year in India involving 67 recorded cases; of these, 28 died (86).

Kentucky Coffee Tree

Gymnocladus dioica
(L.) K. Koch
Leguminoseae

DESCRIPTION: Trees to 30 m, with deeply fissured, scaly bark.

LEAVES: Alternate, deciduous, twice pinnately compound, with 5–9 branches, each branch with 4–7 opposing leaflets.

FLOWERS: Staminate in panicles, with 10 stamens; pistillate on separate trees, in branched clusers with a 5-lobed calyx, 5 petals and 1 pistil; May-June.

FRUIT: Dark red-brown legumes, 15–25 cm long, with thick pulp between the large seeds; Fall, persisting into Winter.

OCCURRENCE: Moist woods from New York to South Dakota and south; sometimes cultivated.

TOXICITY: Seeds and fruit contain the quinolizidine alkaloid cytisine (72, 106, 109).

SYMPTOMS: Vomiting, diarrhea, irregular pulse and coma (72, 109).

NOTES: In a 19th-century case, a woman was poisoned after mistaking the fruit pulp for that of honey locust (*Gleditsia triacanthos* L.), which is sometimes eaten by children. The effects, which began after five minutes and lasted for several hours, were described as narcotic (35). The seeds were roasted and ground by early settlers as a coffee substitute (5). In lactating animals, one pathway of excreting the toxin is in the milk (101). It is possible therefore, to be poisoned by drinking such milk.

Golden Chain
Laburnum anagyroides Medic.
Leguminoseae

SYNONYM: *Cytissus laburnum* L.

DESCRIPTION: Deciduous tree or shrub; to 10 m tall, with hairy twigs.

LEAVES: Trifoliate with ovate leaflets; yellow leaves occur on some cultivated forms.

FLOWERS: Yellow, in long, hanging racemes.

FRUIT: Pods, with several dark-brown, kidney-shaped seeds; sometimes persisting through the Winter.

OCCURRENCE: Native to Europe; cultivated throughout North America; sometimes naturalizing in waste places.

TOXICITY: The quinolizidine alkaloid cytisine occurs in all parts, especially the seeds and bark (4, 72, 101, 106, 109, 134).

SYMPTOMS: Burning of the mouth, thirst, nausea, vomiting, dilated pupils, headache, sweating, irregular pulse, muscle cramps, coma, circulatory collapse, respiratory failure and death (4, 101, 109, 134).

NOTES: Ten children were poisoned in separate incidents during the summer months in Edinburgh, Scotland. Common symptoms, which appeared from ½ to 4 hours after ingestion were nausea, vomiting, paleness, drowsiness, accelerated heart rate, dilated pupils, loss of equilibrium, dizziness and incoordination. With gastric lavage and stimulants, all recovered within 24 hours (121). The beans are cooked in the tropics for food (2–4 changes of water). Insufficient boiling may result in poisoning (72). In lactating animals, cytisine is excreted among other ways in milk. Drinking contaminated milk may thus cause poisoning (101).

Chick Pea
Lathyrus spp.
Leguminoseae

DESCRIPTION: Annual or perennial herbs; stems creeping or climbing.
LEAVES: Pinnate, with tendrils or bristles; leaflets 2–18.
FLOWERS: Pea-like, in racemes, each subtended by a small bract, sometimes falling early; sepals regular or irregular; stames diadelphous, 9 and 1; style bearded on the upper side.
FRUIT: Legume pod with 2 or more seeds.
OCCURRENCE: Roadsides, waste areas, wet places; some (such as *L. sativus* L.) are cultivated for their showy flowers.
TOXICITY: The seeds of *L. sativus* L. contain β-N-oxalyl-L-α,β-diaminoproprionic acid (146), a potentially neurotoxic phenol of unknown structure (130); and a water soluble aliphatic amino acid glycoside with a nitrile group (152).
SYMPTOMS: A diet of almost exclusively chick peas will bring on symptoms in 4–8 weeks (101), or in 3–6 months with up to ½ that amount (155): partial or total paralysis, usually legs, but arms as well in serious cases, pains, prickling sensations, hypothesia and cramps may occur or not; these last symptoms will go away if the diet is corrected, but the paralysis is permanent (101).
NOTES: This affliction known as *neuro-lathyrism*, has been recognized since the time of Hippocrates. If eaten in moderation, the peas do not cause problems (92), however they do continue to be a health problem in central India, where poverty and drought have forced people to eat them as a large part of their diet. The problem persists because the plant will survive where other nutritional species cannot (155, 165).

The toxic principle of sweet pea (*L. odoratus* L.) and everlasting pea (*L. pusillus* Elliott) is β-(gamma-L-glutamyl)-aminopropionitrile in the seeds (49, 101, 155). The active portion of the molecule, β-aminopropionitrile, produces a condition known as osteolathyrism in experimental animals, characterized by skeletal deformities (155).

Lupine

Lupinus spp.
Leguminoseae

DESCRIPTION: Annual, biennial or perennial herbs, rarely shrubs.
LEAVES: Palmate, 1–18 leaflets; stipules sometimes falling early.
FLOWERS: Showy, in terminal racemes, pea-like, purple to blue, flesh-pink, white or yellow; stamens 10.
FRUIT: Flattened legume pod with several seeds.
OCCURRENCE: In old fields, margins, open woods; some cultivated.
TOXICITY: Seeds contain quinolizidine alkaloids such as lupinine and anagyrine (97, 109); alkaloids including sparteine and hydroxylupanine have also been reported (142).
SYMPTOMS: Respiratory depression and slowing of the heart (134).
NOTES: Lupine was named after the wolf, *lupus* in Latin, for it was once believed to destroy the soil by wolfing nourishment out of it (48). Extracts from the plant have been used externally for skin disorders (103). Children may eat the pods by confusing them for garden peas. If eaten by cows, the plant will cause bad flavors in milk (129).

Black Locust
Robinia pseudoacacia L.
Leguminoseae

DESCRIPTION: Medium-sizes trees to 18 m.

LEAVES: Alternate, deciduous, pinnately divided; leaflets 7–19, each oval, smooth.

FLOWERS: Very fragrant, in loose hanging racemes; calyx 5-lobed; petals 5, irregular, cream-colored with a yellow spot on the upper ones; stamens 10; pistil 1.

FRUIT: Flattened pods, 7.5–10 cm long, smooth, reddish-brown, 4–8 seeded; Fall.

OCCURRENCE: Now throughout eastern North America in rich woods, margins and urban areas; sometimes cultivated.

TOXICITY: The flowers are nontoxic (106), but inner bark, young leaves and seeds contain robin and phasin, both toxalbumins (106, 109, 142).

SYMPTOMS: Delayed gastroenteric irritation, with fatigue, loss of appetite, weakness, nausea, coldness of extremities, weak and irregular pulse, dilated pupils, bloody vomiting and diarrhea, labored breathing, dryness of the mouth and throat, paleness and shock; recovery or death in a few days (142).

NOTES: In 1887, 32 boys at the Brooklyn Orphan Asylum ate the inner bark of black locust fence posts being stripped in the orphanage yard. Two were severely poisoned, but all recovered (50). The bark and roots have been used medicinally (142). Clammy locust (*R. viscosa* Vent.) is a related southeastern species with similar toxicity (129).

Purple Sesbane, False Poinciana

Sesbania punicea (Cav.) Benth.
Leguminoseae

SYNONYM: *Daubentonia punicea* (Cav.) Benth.

DESCRIPTION: Deciduous shrub or small tree; stems glabrous, to 3.6 m tall.

LEAVES: Alternate, pinnately divided with 6–20 pairs of leaflets; margins smooth.

FLOWERS: Red-orange in drooping axillary clusters; June-September.

FRUIT: Legume pod with partitions between the seeds.

OCCURRENCE: Native to South America; cultivated as an ornamental; naturalized in disturbed areas of the southeastern Coastal Plain and from Florida to Louisiana.

TOXICITY: The seeds contain saponins (49, 109, 128).

SYMPTOMS: Soon after ingestion or up to 24 hours later: vomiting, loss of appetite, nausea, abdominal pain, diarrhea, and severe gastroenteritis; after absorption: weakness, hemolysis of red blood cells, depression, fast, irregular pulse, shallow and rapid respiration, respiratory failure and death (49, 109, 128).

NOTES: A fatal case was reported when the seeds of senna beans [*S. drummondii* (Ryd.) Cory] were eaten by a small boy (49).

Goat's Rue

Tephrosia virginiana (L.) **Persoon**
Leguminoseae

DESCRIPTION: Perennial herb; roots woody; stems hairy, 2–7 dm tall.
LEAVES: Compound with 8–14 pairs of leaflets.
FLOWERS: In racemes; calyx with long soft hairs; standard yellowish-cream colored; wings and keel red; May-early August.
FRUIT: A flat, hairy legume with lentil-like seeds.
OCCURRENCE: Sandy woods and clearings throughout the East; sometimes cultivated.
TOXICITY: Tephrosin and other unknowns (165).
SYMPTOMS: Internal symptoms unspecified in literature reviewed; dermatitis (142, 143).
NOTES: The crushed stems have been used as a fish poison (142). Indians used the roots to treat intestinal worms, bladder problems and chronic coughing (102).

Fava Bean
Vicia faba L.
Leguminoseae

DESCRIPTION: Coarse, erect, annual vine.
LEAVES: Alternate, compound with 2–6 leaflets; tendrils absent.
FLOWERS: Off-white with a purple spot in the middle; 1-several in leaf axils.
FRUIT: Many-seeded legume, to 35 cm long and 2.5 cm wide.
OCCURRENCE: Native to north Africa and southwest Asia; widely cultivated as a cover crop, for forage; grown in some areas of the southern U.S. for consumption.
TOXICITY: The beans contain cyanogenic glycosides that release HCN (109); the pollen is also toxic to certain people (90, 142).
SYMPTOMS: Within a few minutes of inhaling the pollen and 5–24 hours after eating the beans: yawning, headache, dizziness, nausea, vomiting and fever; serious cases may progress with jaundice, severe hemolytic anemia and death (101).
NOTES: *Favism* is produced by an inherited red blood cell enzyme deficiency (glucose-6-phosphate dehydrogenase) as well as low levels of whole blood glutathione (101). Only persons carrying this genetic trait are affected by the beans or pollen (72). It is seen in a small percentage of blacks and some people of Mediterranean origins (109). The frozen or canned beans are eaten by some Italian Americans (72). The condition is more serious in children and most common in boys (72). In one reported case, a 21-year-old man was admitted to a hospital with abdominal distress and jaundice. He admitted to eating fava beans several times in the previous few days. For two days, he had weakness, headache and abdominal distress. After 15 days, recovery proceeded

rapidly. A history of the family revealed most all members to be susceptible to the beans. The attacks usually lasted 3 days, with jaundice, pallor, dark red urine and weakness; recovery was rapid. The side effects were accepted as a penalty for the privilege of eating the beans (85). In New Mexico, the ground beans have been made into a paste and applied to the chest and back to treat pneumonia (102). Vetch (*V. sativa* L.) seeds contain the cyanogenic glycoside vicianin and the neurolathyrism factor β-cyano-L-alanine as well (145, 148). HCN (to 52 mg/100 g of tissue) released during the breakdown of vicianin reacts with cysteine in the presence of an enzyme to form the factor (145).

Wisteria

Wisteria spp.
Leguminoseae

DESCRIPTION: Twining shrubs or climbing, woody vines.

LEAVES: Pinnate; leaflets entire; stipules falling early.

FLOWERS: Large, pea-like, in terminal, pendent racemes, with early falling bracts; calyx bell-shaped; somewhat 2-lipped; petals blue, violet or white; stamens diadelphous, 9 and 1.

FRUIT: Flattened, several-seeded, dehiscent legume.

OCCURRENCE: Native and introduced; along borders, roadsides, in woods, throughout most of the East; also cultivated.

TOXICITY: Seeds, pods and bark contain the glycoside wisterin and a toxic resin (128, 142).

SYMPTOMS: Gastroenteric irritation, chiefly of the gastric mucosa, nausea, repeated vomiting, abdominal pain, diarrhea, dehydration; recovery usually occurs in about 24 hours (101, 109, 142).

NOTES: In one case, an 8-year-old boy was poisoned by eating 2–3 of the seeds. He recovered with treatment in about 48 hours (128). The genus was named by printer Thomas Nuttall in 1818, in honor of Caspar Wistar, a Philadelphia physician and naturalist (48).

Flax
Linum usitatissimum L.
Linaceae

DESCRIPTION: Slender, erect, annual herbs; 4–12 dm tall.

LEAVES: Mostly alternate, small, narrow, entire.

FLOWERS: In racemose or cymose panicles; sepals 5 petals 5, light-blue; persisting for 1 day; April-September.

FRUIT: Dry capsule with 10 or more seeds.

OCCURRENCE: Native to Europe; cultivated since ancient times for linen fibers; sometimes escaping.

TOXICITY: All parts of the young green plant contain the cyanogenic glycoside linamarin and its homolog lotaustralin (up to 5%) (145).

SYMPTOMS: Difficult breathing, paralysis, convulsions and death have been seen in animals (134); the seed and linseed oil may cause dermatitis (165).

NOTES: Immature seeds are the most toxic part of the plant (134). Workers in the manufacture of linseed oil sometimes get dermatitis and allergic reactions from handling linseed cake (171). Medically, the seeds and oil have been used for coughs, lung, digestive and urinary disorders, gallstones and constipation (109, 113).

Lime
Citrus aurantifolia (L.) Swingle
Rutaceae

DESCRIPTION: Shrubs or trees to 8 m; branchlets stout with green spines.
LEAVES: Alternate, evergreen, simple, aromatic.
FLOWERS: White, solitary or in pairs: calyx 4–5 lobed; 4–5, narrow; stamens 20–25, pistil 1.
FRUIT: Several-seeded berry; surface bumpy, skin yellow-green, not easily separated from the sour pulp.
OCCURRENCE: Native to the tropics: extensively cultivated in south Florida; escaped in some coastal hammocks.
TOXICITY: Fresh lime oil contains terpene hydrocarbons such as d-limonene, α & β-pinene, camphene, sobinene, terpinolene, α-phellandrene, γ-terpinene, p-cymene and others (108).
SYMPTOMS: Photodermatitis, respiratory irritation.
NOTES: Lime harvesters and others frequently handling the fruit sometimes develop dermatitis. Upon exposure to the sun, the skin then becomes itchy and turns brown (128). Cosmetics with lime oil may also cause skin irritation in some people. Volatile oils from the blossoms and fruit peel may cause respiratory irritation in some people (128). If eaten in quantity, sour orange peel (*C. aurantium* L.) is said to cause violent colic, convulsions and even death in children (109).

Trifoliate Orange
Poncirus trifoliate (L.) Raf.
Rutaceae

DESCRIPTION: Shrubs or small trees to 7 m tall. branches still, with stout thorns.
LEAVES: Alternate, deciduous, trifoliate, aromatic, with rounded teeth along the margin, leathery.
FLOWERS: White, calyx small with 5 lobes; petals 5, thin; stamens 8–10; pistil 1.
FRUIT: Rounded, yellow, thin-skinned with sour pulp; seeds white to brown.
OCCURRENCE: Introduced from China; widely cultivated as an ornamental and sometimes escaped; Virginia, south and west.
TOXICITY: The fruit contains an oil, an acrid principle and saponins (106, 142).
SYMPTOMS: Severe gastroenteritis (106); prolonged exposure to the skin to the fruit may cause dermatitis (142).
NOTES: The toxins modify tissue permeability and cause cellular damage (106).

Rue
Ruta graveolens L.
Rutaceae

DESCRIPTION: Small, hardy perennial; base woody; to 1 m.

LEAVES: Deeply dissected, appearing compound, margins entire, toothed, fragrant.

FLOWERS: In terminal corymbs or panicles; sepals 4–5; petals yellow, 4–5; stamens 8–10; May-August.

FRUIT: Capsules with 4–5 locules.

OCCURRENCE: Cultivated in herb gardens; sometimes escaping to waste grounds; North Carolina, Virginia and West Virginia.

TOXICITY: All parts contain furocoumarins (109, 165); the plant also contains rutin (about 2%), a volatile oil, and alkaloids including γ-fagarine, orinine, kokusaginine, skimmianine and graveloine; coumarin derivatives such as bergapten, xanthotoxin, rutamarin, psoralen, isoimperatorin, pangelin, rutarin and other (108).

SYMPTOMS: The leaves cause skin dermatitis (106); subsequent exposure to the sun may lead to photodermatitis (109, 128, 165, 171); internally, large amounts of rue oil cause stomach pain, vomiting, exhaustion, confusion and convulsions; it may be fatal (108).

NOTES: Hippocrates (460–377 B.C.E.) recommended rue for its medicinal properties, and Pliny the Roman writer (23–79 C.E.) said that artisans such as painters and engravers ate it to improve their eyesight (113). In the 16th and 17th centuries, judges took the plant into court to protect against fevers that prisoners might bring in. A tea from it has been used for uterine congestion, stomach ailments, convulsions and hysteria (116). It has also been used for rheumatism, gout and intestinal worms. Finally, the gray leaves served as the model for the suit of clubs in playing cards (113).

Chinaberry

***Melia azedarach* L.**
Meliaceae

DESCRIPTION: Fast-growing, short-lived, small tree to 15 m; bark thin, becoming furrowed with age.

LEAVES: Alternate, deciduous, bipinnate, with many leaflets with toothed margins.

FLOWERS: Many, in large, axillary panicles; sepals 5; petals 5, purplish; stamens usually 10; pistil 1; Spring.

FRUIT: Yellow, 1-seeded drupe, persisting in Winter.

OCCURRENCE: Native to the Himalayas and eastern Asia; widely planted in the South; escaping to thickets, old fields and disturbed areas.

TOXICITY: All parts contain a saponin and a bitter principle; leaves contain paraisine; fruit contains azaridine, a resin; bark has margosine and tannin (142).

SYMPTOMS: Overdoses of an herbal leaf medicine has caused burning of the mouth, scant urine, bloody vomiting, lethargy and sometimes death (142); the fruit has caused nausea, vomiting, constipation or bloody diarrhea, labored breathing, paralysis, convulsions, liver and kidney damage, respiratory failure and death in 12–24 hours (49, 101, 142).

NOTES: A tea from the leaves has caused poisoning in children (72). A young child died from eating 6–8 of the fruits (49, 72, 128, 142). The rootbark has been used as a purgative, an emetic and to help bring on menstruation. The seeds and fruit oil have been used for intestinal worms (113).

Tung Oil Tree
Aleurites fordii Hemsl.
Euphorbiaceae

DESCRIPTION: Small tree with stout branches.
LEAVES: Alternate, simple, deciduous, palmately veined, to 25 cm long.
FLOWERS: Pinkish, in terminal cymes; sepals 2–3, valvate; petals 5, hairy; stamens 8–10; Spring.
FRUIT: Pendent, 3–7 seeded, brown drupe; September.
OCCURRENCE: Introduced from China; cultivated as an ornamental and commercially for the oil in the Southeast; now North Carolina and south.
TOXICITY: The insoluble seed fraction contains a phytotoxin (106, 109, 128, 142) and the irritant ester 16-hydroxyphorbol-12-hexadecanoate-13-acetate (55).
SYMPTOMS: After about ½ hour: severe gastroenteritis, nausea, vomiting, abdominal cramps, diarrhea and weakness; severe cases show dehydration diminished reflexes (72, 109, 128); dermatitis (106, 142).
NOTES: In one case of poisoning, a university student ate one tung nut after mistaking it for a Brazil nut. An hour later, he was weak, dizzy and had stomach pains. Vomiting and diarrhea then reoccurred at short intervals. Drinking fluids to quench an intense thirst only seemed to aggravate the condition, which lasted about 7 hours. The next day he had only a headache and was completely recovered by the third day (52). In another case, a 3-year-old girl ingested an unknown amount of tung nuts. Early symptoms were nausea, vomiting, headache, diarrhea, abdominal cramps and lumbar pain. When she received medical attention about 7.5 hours later, the areas around her ears and lips were blue, breathing was irregular, heartbeat was rapid but regular, pupils were dilated but reactive and reflexes were absent. With treatment, she completely recovered and was discharged the next morning (16).

Spurge Nettle

Cnidoscolus stimulosus (Michx.) Engelm. & Gray
Euphorbiaceae

DESCRIPTION: Monoecious, perennial herb covered with stinging hairs.
LEAVES: Alternate, 3–5 lobed with palmate veins.
FLOWERS: A terminal, compound, dichasial cyme; calyx white, petal-like; petals none; stamens 10–30; March-September.
FRUIT: Capsule with 3 seeds.
OCCURRENCE: Native in sandy woods, along roadsides, in fields, Virginia and south, west to Texas.
TOXICITY: Stinging hairs on the leaves and stems release an irritant chemical when touched.
SYMPTOMS: Painful inflammation; severe reaction in some people but usually lasting no longer than one hour.
NOTES: *Laportea* and *Urtica* have similar stinging hairs. Fainting has been reported in severe cases (128).

Woolly Croton, Hogwort

Croton capitatus **Michx.**
Euphorbiaceae

DESCRIPTION: Monoecious, woolly-haired, annual herb; stems branched, 4–12 dm tall.

LEAVES: Alternate, simple, lanceolate to egg-shaped, margins usually smooth.

FLOWERS: Small, in short racemes, receptacles hairy; staminate with 5 sepals, 5 glands alternating with the petals, and 10–13 stamens; pistillate with 7–12 sepals, petals none, styles 3; June-October.

FRUIT: A 3-seeded capsule.

OCCURRENCE: Sandy soils; New Jersey to Missouri and south.

TOXICITY: Seeds, leaves and stems contain croton oil (109, 142). Phorbol (an ester of the tigliane polyol) has been isolated from the seeds of *C. tiglium* L. Other irritant esters include: 12-0-tetradecanoyl-phorbol-13-acetate and 12-0-acetylphorbol-13-decanoate (55).

SYMPTOMS: Externally the oil causes blistering of the skin (109, 138, 142); internally it has caused gastroenteritis (109).

NOTES: Ten drops of croton oil have been fatal to a dog (129). *C. tiglium* L. oil has caused gastroenteritis and death in humans (109).

Spurge

Euphorbia spp.
Euphorbiaceae

E. splendens

DESCRIPTION: Erect or prostrate herbs or shrubs, some cactus-like, most with milky, irritant sap.
LEAVES: Alternate or opposite, simple, entire or toothed.
FLOWERS: A cup-like structure (cyathium) contains several very small staminate flowers and usually 1 pistillate flower that is pedicelled; 4–5 petal-like glands occur around the rim.
FRUIT: Globose, pedicelled capsule with 3, 1-seeded locules.
OCCURRENCE: Native and introduced; cosmopolitan in borders, fields, margins, open woods, parking lots and sidewalk crevices; some cultivated.
TOXICITY: The irritant sap contains various principles.
SYMPTOMS: The milky sap irritates the eyes, mouth and gastrointestinal tract (72, 101, 142); it causes dermatitis in some people (72, 142).
NOTES: In one case of fatal poisoning, a young woman died after using snow-on-the-mountain (*E. marginata* Pursh) to induce abortion (72). Poisoning has resulted from cooking caper spurge (*E. lathyris* L.) after mistaking them for true capers (*Capparis spinosa* L.) (134). Experience has shown that poinsettia (*E. pulcherrima* Willd. ex Klotzsch) normally causes nothing more than minor skin irritation and probably does not deserve the reputation of being very toxic (117). The spiny candelabra cactus (*E. lactea* Haw.) often causes severe skin and eye irritation when pruned (128). Pencil tree (*E. tirucalli* L.) with acrid sap, is sold as a potted plant. Beggars of the past rubbed spurge leaves on the legs and feet to cause sores and elicit pity from passersby (93). If eaten by cows, the plants cause a bad flavor in milk (129).

Spurges continued on page 100.

Spurge

E. lactea

E. heterophylla

E. marginata

E. tirvcalli

Manchineel

Hippomane mancinella L.
Euphorbiaceae

DESCRIPTION: Small trees to 12 m tall; with milky juice, bark warty and fissured.
LEAVES: Opposite, simple, ovate to elliptical, entire or toothed, leaf stalks with 1–2 glands.
FLOWERS: Unisexual on the same or separate trees; yellow or red.
FRUIT: Light-green to yellow drupes.
OCCURRENCE: Native in coastal hammocks of south Florida; rare except in the Everglades National Park.
TOXICITY: The fruit and milky sap contain hippomanin A and B (128); the irritant esters huratoxin and mancinellin (2, 55).
SYMPTOMS: Occurring 1–2 hours after ingesting the fruit or leaves: swelling and blistering of the lips, mouth and throat, bloody diarrhea, vomiting, abdominal pain, shock and occasional death in severe, untreated cases (72, 102, 142). After about 1/2 hour, the milky sap will cause severe skin dermatitis (72, 101, 106, 109, 128, 135, 142, 165) and even temporary blindness if rubbed in the eyes (101). Rain dripping from the leaves contains enough toxin to cause dermatitis (128).
NOTES: Indians used the sap in medicine, as an arrow poison and to poison the water supply of enemies (128). Smoke from the burning plant causes eye irritation and headache (128, 142).

Sandbox Tree
Hura crepitans L.
Euphorbiaceae

DESCRIPTION: Large, spiny tree with clear, sticky sap; to 20 m tall.
LEAVES: Simple 6–20 cm long; downy beneath.
FLOWERS: Male in an erect spike, bright red; female appearing like a small mushroom.
FRUIT: Ripe capsules spring open loudly and throw the seeds for a great distance.
OCCURRENCE: Native in the American tropics; occasionally planted in south Florida as an ornamental.
TOXICITY: The irritant and carcinogenic daphnane ester huratoxin in unspecified parts (55, 109, 153); all parts with the toxalbumin hurin (128, 142).
SYMPTOMS: Eating 2–3 seeds that taste pleasant will cause dim vision, accelerated pulse, abdominal pain, nausea, vomiting, bloody diarrhea, delirium and sometimes death; the sap may cause severe skin irritation with blisters (101, 106, 109, 127, 135, 142, 165) as well as temporary blindness if contacting the eye (128, 142). Smoke from the burning wood is an eye and respiratory irritant (128, 142).
NOTES: The seeds have been used to kill fish and undesirable animals in some countries (142). Dust from sawing or sanding the wood is irritating to the eyes and nose (128).

Physic Nut

Jatropha curcas L.
Euphorbiaceae

DESCRIPTION: Small tree or shrub; to 2.5 m tall.
LEAVES: Alternate, 3–5 lobed, palmately veined.
FLOWERS: Small, greenish-yellow, with hairs; unisexual.
FRUIT: Yellow, 3-seeded capsule, about 3 cm in diameter.
OCCURRENCE: Native and cultivated in south Florida.
TOXICITY: Seeds and sap contain jatrophin (curcin) a toxalbumin and a powerful purgative oil; roots and leaves are also toxic (106, 109, 142).
SYMPTOMS: After a few minutes to several hours: burning in the throat, nausea, abdominal pain, vomiting, bloody diarrhea and coma; severe cases may show muscle spasms, labored breathing, pupil dilation, dehydration, collapse and even death (72, 101, 109, 128, 142).
NOTES: Adults are usually poisoned by overdoses of the purgative oil and children by the pleasant tasting seeds (101, 109). In September 1958, a 3-year-old boy was admitted to a hospital in Honolulu, after having eaten about 10 of the seeds in a neighbor's yard. Early symptoms soon after dinner included vomiting and diarrhea with particles of the seeds in both. He was weak, lethargic, cyanotic and slightly dehydrated. With treatment, he was discharged after two days, completely recovered (80).

Castor Bean
Ricinus communis L.
Euphorbiaceae

DESCRIPTION: Shrub-like, monoecious herb; stems red to green, to 3.6 m tall.

LEAVES: Alternate, simple, greenish-red with 5–11 palmate lobes, toothed.

FLOWERS: In terminal racemes; pistillate above; staminate below; sepals 5, valvate; petals none; stamens many; July-October.

FRUIT: Greenish-red, oval, fleshy, spined capsule; seeds 3, black, white or mottled.

OCCURRENCE: Native to Africa and Asia; cultivated and sometimes escaping in the southern U.S.; grown commercially for castor oil.

TOXICITY: All parts, especially the seeds contain ricin, a phytotoxin; ricinine, an alkaloid; HCN, a powerful allergen and unknowns (108, 142, 163).

SYMPTOMS: Nausea, vomiting, headache, colic, general malaise, bloody diarrhea, dehydration, thirst, burning of the throat, fever, loss of consciousness, cyanosis, tachycardia, convulsions, liver and kidney damage, proteinuria and a rise in excretion of non-protein nitrogen (15, 139); volatile emanations especially when in bloom sometimes cause allergic respiratory irritation (106, 126, 128); allergic responses from handling the leaves sometimes result in eye irritation, bronchial asthma and contact dermatitis (106, 142).

NOTES: Ricin is a large protein molecule. Like other proteins, it is denatured by heat and thus rendered nontoxic (165). The allergen, however, is heat stable (128). One to several seeds may be lethal if chewed and swallowed (139). Poisoning is unlikely though, if swallowed without breaking the hard seedcoat. This plant is the source of castor oil, a very effective purgative in little use today (109). Ricin is not present in the oil, but remains in the seed-cake (101). The seeds have also been used as an oral contraceptive in Algeria (25), a very dangerous practice. The high toxicity has been known since ancient times and hundreds of cases of human poisoning have been described (15). Some people are allergic to the pollen. Such was the case in May 1961, when a woman in Ramle, Israel, sought medical relief from sneezing, inflammation of nasal mucous membranes, itching, lacrimation, coughing and wheezing. The symptoms began in early spring and lasted until May. A visit to her home revealed many castor bean plants growing nearby. Tests confirmed her allergy to the pollen which is airborne at that time of the year (111).

Popcorn Tree
***Sapium sebiferum* (L.) Roxb.**
Euphorbiaceae

DESCRIPTION: Fast-growing, medium-sized tree to 15 m; bark smooth, becoming fissured with age.
LEAVES: Alternate, simple, widest near the base, entire, but with wavy margins, 2 glands at blade base.
FLOWERS: In long spikes to 10 cm; staminate above, pistillate below; Spring.
FRUIT: Rounded, 3-lobed capsule, 1.2–1.8 cm in diameter, falling away and leaving white seeds; Fall.
OCCURRENCE: Native to China and Japan; introduced as an ornamental; now wild in the Coastal Plain, South Carolina to Texas.
TOXICITY: Unripe berries and plant sap contain unidentified toxins.
SYMPTOMS: Gastrointestinal upset with nausea and vomiting (49); skin dermatitis (19).
NOTES: The fruit, leaves and root bark have been used medicinally (142), and eating the ripe, white-hulled seeds has not caused problems (49). The latex, however, is toxic and has been used as an arrow poison in central America (109). In one case, two children in Marietta, Georgia, experienced contact dermatitis with itching and redness after playing in a clump of the saplings (19). The stems, with white, exposed seeds still attached, are used in dried floral arrangements and sold extensively by the "flower ladies" of the historic Market District of Charleston, South Carolina.

Boxwood

Buxus sempervirens L.
Buxaceae

DESCRIPTION: Monoecious, evergreen shrub; stems angular or winged.

LEAVES: Opposite, simple, leathery, pale beneath, dark green above, midrib whitish, entire.

FLOWERS: Small, in axillary clusters; staminate lateral, sepals 4, stamens 4; pistillate terminal, sepals 6, ovary with 3 cells.

FRUIT: Three-celled capsule with black seeds.

OCCURRENCE: Native to Europe, western Asia and northern Africa; cultivated throughout the U.S.

TOXICITY: The alkaloid buxine is found in the leaves, twigs and roots (72, 109, 142).

SYMPTOMS: Vomiting and diarrhea; large amounts may lead to convulsions, respiratory failure and death (72, 109); the leaves may cause dermatitis (134).

NOTES: The plant has been used as a purgative (113), but animals have died from browsing the leaves (113, 142).

Mango
Mangifera indica L.
Anacardiaceae

DESCRIPTION: Tree to 15 m; bark thick, corky, gray.

LEAVES: Alternate, evergreen, simple, crowded together almost in a whorl, leathery, dark-green, smooth.

FLOWERS: Fragrant, in large, branched clusters; bi- and unisexual on the same or different trees; calyx 4–5 lobed, petals 4–5, yellowish-green to white; stamens 1–2 fertile, 3–4 sterile; pistil 1.

FRUIT: Large, rough to oval-shaped, up to 2 kg, reddish-pink with tinges of yellow or green; skin thin, covering an edible pulp with 1 stone-like seed.

OCCURRENCE: Native to India; cultivated in south Florida for the delicious fruit; escaped, now wild in hammocks.

TOXICITY: Urushiol and oleoresins occur in the fruit peel and sap of the tree (109).

SYMPTOMS: Severe dermatitis that may lead to vesicular eruptions (72, 106, 127, 135, 142, 165); all parts except the ripe fruit may be toxic internally (142); airborne odors from the flowers may cause respiratory irritation (128, 142); the pollen may cause dermatitis (109).

NOTES: Smoke from the burning wood causes eye and skin irritation (128, 142). In one case, a woman experienced vesicular eruptions on her face and lips on several occasions after eating mangos. Skin tests with the pulp were negative, but tests with the peel reacted positively (26). The fruit peel should be removed before being eaten.

Poisonwood

Metopium toxiferum
(L.) Krug & Urban
Anacardiaceae

DESCRIPTION: Shrub or tree to 12 m; bark scaly, reddish-brown.

LEAVES: Opposite, pinnate, leaflets 3–7, tip rounded and often notched, entire, leathery, shiny.

FLOWERS: Very small, in axillary panicles, yellowish-green; unisexual on separate trees; calyx 5-lobed; petals 5; stamens 5 in staminate flowers; pistil 1 in pistillate flowers.

FRUIT: Drupes, yellowish-orange, shiny, 1-seeded.

OCCURRENCE: Native to south Florida; in hammocks, pinelands and on dunes; cultivated.

TOXICITY: The sensitizer is variously given as an alkyl catechol (66), a mono- or di-hydric phenol (142), and urushiol (109).

SYMPTOMS: All parts cause allergic contact dermatitis (72, 106, 127, 128, 138, 142, 165); fever and other internal complications may also occur (128, 142); the sap may leave black stains on the skin (128, 142).

NOTES: The sap blackens and hardens upon exposure to air (101). Smoke from the burning tree is very irritating and may even cause temporary blindness (128).

Brazilian Pepper, Florida Holly

Schinus terebinthifolius Raddi
Anacardiaceae

DESCRIPTION: Low-branching, bushy tree; to 9 m tall.
LEAVES: Compound; aromatic; midribs reddish; Leaflets 5–9.
FLOWERS: Small, white, in dense clusters; Unisexual, on separate trees; mostly October–November.
FRUIT: Small, red, round, berry-like; with yellow seeds.
OCCURRENCE: Introduced from South America as an ornamental tree; it was spread throughout south Florida by birds and is now a fast-growing, noxious species.
TOXICITY: All parts contain urushiol and other principles (109).
SYMPTOMS: Contact with the resin by trimming the tree causes a stinging rash, which may include eye inflammation and facial swelling (128, 142). Volatile emanations from the blossoms or crushed fruit causes respiratory distress in some people (127, 128, 142). Eaten in quantity, berries and sap cause gastroenteritis and vomiting (128, 142).

Poison Ivy, Oak and Sumac

***Toxicodendron* spp.**
Anacardiaceae

Poison oak

Poison ivy

The following species occur in the East; characteristics that identify and separate them are:

1. *T. radicans* (L.) Kuntze (poison ivy)
2. *T. quercifolium* Michx. Greene (poison oak)
3. *T. vernix* (L.) Kuntze (poison sumac)

DESCRIPTION: 1. Climbing or trailing vine. 2. Shrub, never climbing. 3. Shrub to small tree.
LEAVES: 1,2. Alternate, trifoliate, glabrous above, hairy beneath. 3. Alternate, pinnately divided, with 7–13 leaflets.
FLOWERS: 1,2,3. Small, axillary, pendent, in clusters; unisexual, on separate plants; early Summer.
FRUIT: 1. Yellow, glabrous to hairy drupes; August-November. 2. Yellow, hairy drupes; August-November. 3. Cream-colored, glabrous drupes; August-November.
OCCURRENCE: 1. In disturbed areas, margins, flood plains; throughout the East. 2. Sandy soils, pine woods, old fields; New Jersey and south. 3. Swampy areas, pine barrens; throughout the East.
TOXICITY: All parts contain various forms of the irritant phenolic substance 3-pentadecylcatechol, commonly known as urushiol or toxicodendrol (14, 49, 72, 101, 109, 142).

Poison Ivy, Oak and Sumac

SYMPTOMS: Allergic contact dermatitis with itching, redness, swelling and blisters (72); severe cases may require hospitalization; eating the fruits causes gastrointestinal irritation (142) and has caused death (49).

NOTES: The sap, which turns black on exposure to air, was one of the few natural sources of that color before the introduction of synthetic dyes (159). Medically, the plant has been used to cure eczema and shingles and by Indians of southern California to cure ringworm (159). The irritant sap is spread when the plant is touched by animals, clothes or tools. Droplets of it may also be carried by dust and ash particles in smoke when the plant is burned (101). In commonly used areas, the plant should be removed manually or with chemical herbicides.

Poison sumac

Holly

Ilex spp.
Aquifoliaceae

I. glabra

I. opaca

I. vomitoria

DESCRIPTION: Trees or shrubs.

LEAVES: Alternate, simple, evergreen or deciduous.

FLOWERS: Very small, greenish-white, male and female on separate trees; calyx 4–6 lobed; petals 4–6, free to somewhat fused at the base; stamens 4–6 in male flowers; pistil 1, ovary with 4–8 cells.

FRUIT: Berries of various colors.

OCCURRENCE: Native and introduced; some cultivated.

TOXICITY: Berries contain the bitter principle illicin (49).

SYMPTOMS: Consuming large quantities may cause vomiting and purging (109).

NOTES: Yaupon (*I. vomitoria* Soland. in Ait.), a Southeast coastal species, was used in a strong drink by Indians (72). The leaves were used to promote vomiting (109). North Carolina Indians drank an infusion from the berries of *I. opaca* Soland. in Ait. as a cardiac stimulant (109). The bark and fruits were also used in the 19th century as a laxative, for parasitic worms, coughs, fever and other things (102). The foliage and berries should be safeguarded at Christmas to avoid possible poisonings of children.

Strawberry Bush, Eastern Wahoo

Euonymous atropurpureus Jacq.
Celastraceae

DESCRIPTION: Small shrub or tree to 8 m.

LEAVES: Opposite, simple, deciduous, finely toothed, somewhat hairy beneath.

FLOWERS: Six-18 in branched, axillary clusters; calyx 4-lobed; petals 4; stamens 4; pistil 1; Spring.

FRUIT: Dehiscent, 4-lobed capsule, reddish-purple; Fall.

OCCURRENCE: Thickets, borders, ravines and moist woods; throughout most of the East; may be cultivated.

TOXICITY: Bark, leaves and fruit contain unknown toxins; spindle tree (*E. europeus* L.) a cultivated species, contains the glycosides erobioside, evomonoside and evonoside (4).

SYMPTOMS: Weakness, chills, vomiting, diarrhea followed by convulsions and unconsciousness (72, 109,142).

NOTES: This plant was a popular diuretic in the 19th century. It was also used for bronchial congestion, fever and indigestion (113). Pioneers used the bark as a laxative and liver stimulant. An oil from the seed was used for head lice (102).

Buckeye, Horse Chestnut

Aesculus spp.
Hippocastanaceae

DESCRIPTION: Small to medium-sized trees or shrubs.
LEAVES: Opposite, deciduous, palmately compound; leaflets 5–7, toothed, with long stalks.
FLOWERS: In clusters at the ends of branches; calyx tubular or bell-shaped, 5-lobed; petals 4–5 unequal, 8 or 10; pistil 1.
FRUIT: Capsules, smooth to spiny, with 1–3 large, shiny, brown nuts.
OCCURRENCE: Rich, moist woods and thickets; some cultivated.
TOXICITY: Aesculin, a hydroxy derivative of coumarin has been found in the leaves, bark, young twigs and seeds of several species (72, 101, 109).
SYMPTOMS: Mucous membrane inflammation, nausea, headache, salivation, fever, abdominal pain, thirst, vomiting, diarrhea, depression, stupor, incoordination, weakness, dilated pupils, convulsions, circulatory and respiratory failure; possibly death (49, 72, 142).
NOTES: The leaves, fruit and bark have been used as an expectorant, and an astringent in reducing mucous congestion (113), but the raw nuts have been fatal to children (101). Indians ate the ripe nuts after thoroughly roasting them (129). European herbalists recommended the fruit be kept in one's pocket to prevent and cure arthritis (113). People in Appalachia have done this also (102). In one case of poisoning, a 4.5-year-old boy became ill one afternoon and was restless throughout the night. The next day he admitted eating horse chestnuts with two play-

Buckeye, Horse Chestnut

mates. They had experienced headache, vomiting and fever. He soon recovered, but ate the nuts again while on a walk. That night, he was restless and the following morning incoordinated, drowsy, without appetite and salivated heavily during meals. On the morning after that, he became unconscious and was hospitalized with paralysis of the left side of the face and progressive respiratory paralysis, which led to his death about 48 hours after the second ingestion (157).

Akee
Blighia sapida Koenig
Sapindaceae

DESCRIPTION: Stiff-branched tree, to 30 m tall.
LEAVES: Pinnately divided, to 30 cm long; leaflets smooth, to 15 cm long.
FLOWERS: In drooping spikes, small, white, fragrant.
FRUIT: Straw-colored to red capsules, with 3 cells.
OCCURRENCE: Native to west Africa; grown in south Florida for the edible fruits.
TOXICITY: Immature or spoiled fruit aril contains hypoglycin A (101, 142); seeds contain this and its glutamyl derivative hypoglycin B, and are toxic at all times (101, 128).
SYMPTOMS: After several hours: nausea, vomiting, then a period of drowsiness; after a few more hours severe vomiting, convulsions, coma and death. The blood pH becomes altered and blood sugar decreases (severe hypoglycemia) (72, 87, 101, 109, 142).
NOTES: The fruit aril is edible raw or cooked. They are even canned for export from Jamaica, but "vomiting sickness" as it is called, has caused many deaths there from improper fruit selection and/or preparation (101, 128, 142). Circumstantial evidence pointed to akee as the cause of death of a previously healthy young boy in Pueblo Nuevo, Panama, in 1942. For breakfast, he had only tea and was later seen by neighbors playing under an akee tree with fruits on the ground while his family was away at market. He vomited that afternoon, refused supper, and went to bed at 8 P.M. During the night, he awoke crying but was comforted back to sleep. At 2:30 A.M., he awoke, cried deliriously, kicked his legs and died. An autopsy revealed no evidence of chronic disease (95).

Buckthorn
Rhamnus spp.
Rhamnaceae

DESCRIPTION: Small trees or shrubs; branches sometimes rigid, spine-tipped.
LEAVES: Simple, alternate, deciduous to evergreen, entire or toothed.
FLOWERS: Male and female or perfect, all on the same tree; 4–5 parted.
FRUIT: Berry-like, fleshy, 2–4 seeded.
OCCURRENCE: Native and introduced; moist woods of the Piedmont; some cultivated.
TOXICITY: *R. catharticus* L. is known to contain the glycoside rhamnoemodine, glucosidorhamnoside, rhamnocathardin and shestrin; unripe fruits contain saponins (4).
SYMPTOMS: Gastroenteric irritation chiefly of the intestinal mucosa, purging and collapse (106, 109).
NOTES: Buckthorn is one of the thorny shrubs sometimes supposed to have composed Christ's crown at the crucifixion (103). *R. purshiana* DC. of the West Coast is used in making the laxative cascara (72). At one time a syrup taken from the fruit was used as a purgative but is now generally judged too violent in action (134). It was deleted from the British Pharmacopoeia in 1934 (93). The unripe berries are a source of yellow dye (142). Coyotillo [*Karwinskia humboldtiana* (J. A. Schultes) Zuccar.] is another poisonous shrub of the buckthorn family that grows in the Southwest. Used chronically, the fruit has caused paralysis. In one case, a 5-year-old girl started limping and fell frequently because her legs gave away. She was admitted to a hospital a month later with complete limb paralysis and respiratory difficulties complicated by

pneumonia. The case history revealed that the child and her siblings had been eating unknown amounts of coyotillo fruit for several weeks before the symptoms began. Her condition slowly improved, and within a month of admission she could sit, hold her head up and raise her arms and legs (30).

Virginia Creeper
Parthenocissus quinquefolia (L.) Planchon
Vitaceae

DESCRIPTION: High climbing vine; pith white; tendrils branched with adhesive disks.
LEAVES: Alternate, palmately divided, leaflets 5, elliptical, toothed.
FLOWERS: In panicle-like cymes; petals 5, yellowish-green; stamens 5; May-August.
FRUIT: Drupes with 1–3 seeds.
OCCURRENCE: Common in woods, disturbed areas and margins; throughout the East; sometimes grown on walls and fences.
TOXICITY: Oxalic acid (106); the berries are usually the cause of poisoning, but leaves may be toxic as well (49, 72, 106, 142).
SYMPTOMS: After about 24 hours; nausea, abdominal pain, bloody vomiting and diarrhea, dilated pupils, headache, sweating, weak pulse, drowsiness, electrolyte imbalance, cramps and facial muscle twitching, kidney damage and collapse; acetone is found in the urine and on the patient's breath (49, 142).
NOTES: In a few cases, the berries are suspected as the cause of lethal poisoning in children (101, 109).

Mastwood

Calophyllum inophyllum L.
Hypericaceae

DESCRIPTION: Erect, dense tree; to 15 m tall.
LEAVES: Leathery, smooth, to 15 cm long.
FLOWERS: Aromatic, with 4 white petals.
FRUIT: Yellow, thin-shelled, with a large white kernel.
OCCURRENCE: Native to tropical Asia and the Pacific islands; commonly planted in Florida as an ornamental tree.
TOXICITY: Leaves contain HCN and a saponin; domba oil from the seeds contains calophyllic acid, inophyllic acid and calophyllolide; the fruit may also be toxic (128, 142).
SYMPTOMS: Eaten raw, the seeds cause severe vomiting (142); the sap is irritating to the skin and eyes (128).
NOTES: The half-ripe seed endosperm is eaten after being pickled (127), but poisonings have occurred among native Indians (142). The seed oil is used in Old World tropics to treat warts. Lepers there help relieve their pain by intramuscular injections of the refined oil (109). The plant sap has also been used as an arrow poison (142).

Wicopy, Leatherwood

Dirca palustris L.
Thymelaceae

DESCRIPTION: Shrubs up to 18 dm tall.
LEAVES: Simple, elliptical, entire, deciduous; 5–8 cm long.
FLOWERS: Two-3 in axillary clusters, with or just before the leaves appear; floral tube yellow; stamens 8, inserted, unequal in size; ovary with 1 locule; March-April.
FRUIT: A red drupe; June-July.
OCCURRENCE: In moist woods, along creeks; Canada south to Florida and Oklahoma; may be cultivated.
TOXICITY: All parts contain an irritant resin (106, 142).
SYMPTOMS: Gastrointestinal irritation with vomiting and diarrhea (106, 142); leaves and bark may cause skin irritation with blisters (72, 106, 142, 165).

Eucalyptus
Eucalyptus spp.
Myrtaceae

Eucalyptus St. John

Eucalyptus tonquata

DESCRIPTION: Trees that may become very tall.

LEAVES: Stiff, alternate except sometimes opposite in young shoots; simple, entire.

FLOWERS: Solitary, in clusters; top- or bell-shaped, sepals and petals 4; stamens many.

FRUIT: Capsules opening by many valves; seeds many, small.

OCCURRENCE: Introduced and cultivated.

TOXICITY: The leaves of all species contain oil of eucalyptus and cyanogenic glycosides that release HCN (109); the oil contains 70–85% eucalyptol (1,8-cineole) and monoterpene hydrocarbons such as α-pinene, d-limonene, p-cymene and camphene (108).

SYMPTOMS: Gastroenteritis, nausea, vomiting, diarrhea, labored breathing, stupor, paralysis, convulsions and death which may be due to respiratory failure (142); dermatitis (106, 142); the pollen may induce or aggravate bronchial asthma (109).

NOTES: The oil has been used internally and externally in medicine as well as an insecticide (142).

Bottlebrush Tree, Cajeput

Melaleuca quinquenervia (Cav.) S. T. Blake
Myrtaceae

DESCRIPTION: Trees up to 15 m; branches hanging, bark shedding in strips to show the inner red bark.
LEAVES: Alternate, simple, broadest near the middle, 5–10 cm long, leathery, glandular, aromatic when crushed.
FLOWERS: Showy, white, produced around branchlets; giving a bottlebrush appearance.
FRUIT: Dehiscent capsules, small, brown, with many small seeds.
OCCURRENCE: Introduced and planted in south Florida; escaped in low wet areas and cypress swamps.
TOXICITY: Cajeput oil, which contains 14–65% cineole, pinene and terpinol occurs in the leaves and twigs (108).
SYMPTOMS: Volatile emanations from the blossoms cause respiratory irritation, facial rash, nausea and headache (127, 128, 142); the bark may cause dermatitis (142).
NOTES: The oil is used medicinally, but large amounts may be toxic with kidney and gastrointestinal irritation (128, 142). Externally, it has been used to treat scabies and other parasitic diseases (102). In one case, a woman experienced blisters and itching after scratching her wrists on the broken roots of a tree being dug up (125).

Hercules Club
Aralia spinosa L.
Araliaceae

DESCRIPTION: Shrub or small tree to 7 m; branches spiny.

LEAVES: Alternate, deciduous, bipinnate, at the ends of branches; margins with sharp teeth.

FLOWERS: Very small in terminal clusters; whitish; Summer.

FRUIT: Fleshy berries; Autumn.

OCCURRENCE: In low hammocks, along streams, in moist woods, as far north as New Jersey and west to Missouri and Texas; may be cultivated.

TOXICITY: Possibly aralin, a volatile oil; and a resin (142).

SYMPTOMS: Toxic internally, no details given (72, 142); bark and roots may cause dermatitis with blisters and inflammation (126, 142, 165).

NOTES: The plant has been used medically in small amounts but overdoses are toxic (142). Indians used a root and bark decoction for fever and to purify the blood (102). The berries may be toxic in large amounts (72).

English Ivy
Hedera helix L.
Araliaceae

DESCRIPTION: Evergreen, woody, climbing, creeping vine, with many short, aerial roots.

LEAVES: Alternate, simple, leathery, palmately veined, 3–5 lobed.

FLOWERS: Small, greenish, in terminal umbels.

FRUIT: Black drupes.

OCCURRENCE: Native to Europe; cultivated in the U.S. as a ground and wall cover; sometimes escaping.

TOXICITY: Stems, leaves and fruits contain a saponin and the sapogenin hederagenin and hederin (4, 72, 109, 142).

SYMPTOMS: Anxiety, excitement, nervousness, thirst, headache, fever, dilated pupils, facial rash, nausea, vomiting, salivation, stomach pains, diarrhea, labored breathing, incoordination, convulsions, coma and death (4, 72, 109, 134, 142); severe allergic contact dermatitis may occur in some people after about 48 hours (4, 49, 72, 90, 106, 134, 142, 165).

NOTES: In Greece, ivy is called *cissos* after the nymph Cissos, who danced herself to death before Dionysus the god of wine (113). Children have been poisoned by the fruits (129), but the bitter taste usually discourages ingestion of large amounts (49). The leaf has been used medicinally in treating wounds and ulcerative sores, to combat fleas and for toothache (93). An infusion of the plant has been used for bruises (134). It is believed in some Mediterranean areas that taking 1 gram of the finely powdered fruit will bring about sterility (171).

Aralia
Polyscias spp.
Araliaceae

DESCRIPTION: Shrubs, 6–7.5 m tall; branches, slender, upright.
LEAVES: Compound, variegated with white; leaflets 3–7 toothed; evergreen.
FLOWERS: Very small, white.
OCCURRENCE: Ornamental hedge plants in Florida and the Gulf Coast states.
TOXICITY: The foliage contains saponins and unknowns (128, 142).
SYMPTOMS: Allergic contact dermatitis causing an itching rash with sores and swelling (72, 128, 142, 165); internal irritation if consumed (165).
NOTES: Irritation usually develops when the plants are being trimmed (128). *P. balfouriana* Bailey and *P. guilfoylei* Bailey are two species that occur in south Florida.

Water Hemlock, Spotted Cowbane
Cicuta maculata L.
Apiaceae

DESCRIPTION: Perennial herbs; stems glabrous, purple-striped or mottled, hollow except at nodes, 1–2 m tall; roots tuberous, short, with several chambers.

LEAVES: Alternate, 2–3 pinnately divided, to 6 dm long; leaflets narrow, 2.5–10 cm long, toothed; petioles clasping the stem.

FLOWERS: Small, white, in terminal and lateral, compound umbels; May–September.

FRUIT: Very small, dry, ribbed schizocarps.

OCCURRENCE: Moist thickets, streambanks, swamps, ditches; throughout the eastern U.S. and Canada; may be cultivated.

TOXICITY: This is considered the most violently toxic plant of the northern temperate zone (101). Cicutoxin is an unsaturated alcohol that acts on the central nervous systems in about ½ hour (101, 142). It occurs in the roots and much less in above ground parts (72, 142). Drying does not reduce the toxicity very much (49, 150).

SYMPTOMS: Nausea, salivation, vomiting, diarrhea, abdominal pain, dilated pupils, fever, delirium, convulsions with intermittent periods of calm, complete paralysis, weak and rapid pulse, respiratory or circulatory failure and death (72, 101, 128, 142, 150).

NOTES: There is enough toxin in one rhizome to cause fatality (72, 109, 142). The roots are easily mistaken for those of wild parsnip (*Pastinaca sativa* L.) and have the smell of celery (72, 101, 109). Children have been poisoned using peashooters made from the hollow stems (72, 109). Many cases of poisoning have been reported through the years. One interesting case occurred on June 16, 1925, at the New Haven County

Home in Connecticut. After classes, 17 boys aged 9–13 years ate water hemlock flowers, leaves and rootstock in a swampy area near the playground. They became ill about 5:30 P.M. and 5 were in convulsions when a physician arrived. The others were vomiting and appeared seriously ill. The 5 with convulsions had eaten pieces of the rootstock, while the others ate only leaves or flowers. After gastric lavage, enemas and other symptomatic treatment procedures, all recovered (62). In a more recent case, a 26-year-old man on a fishing trip in eastern Oregon ate a thumb-sized piece of a tuber he thought to be wild carrot. After treatment and recovery, he admitted to frequent sampling of wild plants and considered himself knowledgeable of edible species (32).

Poison Hemlock, Fool's Parsley

***Conium maculatum* L.**
Apiaceae

DESCRIPTION: Biennial herb; stems hollow, purple, spotted or lined, to 2.4 m tall; taproot solid, long, turnip-like.

LEAVES: Large, alternate, 3–4 pinnately divided, leaflets minute; stalk clasping stem.

FLOWERS: In compound, terminal and lateral umbels; petals white; May–August.

FRUIT: Very small, dry, ribbed.

OCCURRENCE: Native to Eurasia; naturalized in waste and marshy areas throughout the U.S.; may be cultivated.

TOXICITY: The 2 main piperidine alkaloids coniine and γ coniceine occur in all parts especially leaves, unipe fruit and roots (49, 97, 101, 106, 109, 163). Others mentioned include N–methyl coniine, conhydrine and pseudoconhydrine (49, 101, 141, 163).

SYMPTOMS: Nervousness, weakness, nausea, vomiting, diarrhea, confusion, pupil dilation, weak pulse, respiratory distress, convulsions, coma, coldness of extremities, respiratory failure and death (49, 101, 109, 163).

NOTES: The coniferous hemlock tree (*Tsuga spp.*) is not poisonous at all (72, 101). Conium resembles carrot except the latter has hairy leaves and stems. The leaves may be mistaken for parsley, the roots for parsnip and the seeds for anise (72, 113, 134). The early Greeks used extracts from this plant as a means of execution and suicide. Socrates was killed by such a potion in 339 B.C. (4, 40, 163). Overdoses of an extract of the unripe fruit used medicinally have caused fatalities (90). If consumed by cows, the plant also causes bad flavors in milk (129). Do not allow children to make toys from the hollow stems.

Wild Carrot, Queen Anne's Lace

Daucus carota L.
Apiaceae

DESCRIPTION: Erect, biennial herb; stems bristly.

LEAVES: Alternate or basal, pinnately divided, segments narrow, hairy.

FLOWERS: White, in flat-topped umbels, subtended by long, finely divided bracts; petals unequal; May–September.

FRUIT: Ribbed, 2-carpellate, with bristly hairs.

OCCURRENCE: Introduced; common in open areas such as meadows and pastures.

TOXICITY: Leaves contain furocoumarins (106, 142).

SYMPTOMS: Some people get allergic contact dermatitis from the leaves especially when wet (129); later exposure to the sun may cause mild photodermatitis (106, 109).

NOTES: Domesticated carrots are vegetatively similar, but have a thick root that is formed during the second year of growth (103). Preparations from the seeds were once used for urinary and menstrual problems (159), and the boiled, mashed roots were applied as a poultice for bruises and cuts (102). The colloquial name was given to honor Queen Anne, wife of James I of England (159).

Wild Parsnip
Pastinaca sativa L.
Apiaceae

DESCRIPTION: Coarse, pubescent biennial; stems ribbed hollow, up 2 m tall; taproot fleshy or fibrous.

LEAVES: Pinnate or bipinnately divided, irregularly dentate, somtimes lobed.

FLOWERS: In compound, lateral and terminal umbels; involucral bracts rarely present; petals yellow; May–October.

FRUIT: Glabrous, elliptic, 5–7 mm long.

OCCURRENCE: Native to Europe; grown for the edible root.

TOXICITY: All parts contain furocoumarins (55, 72, 165).

SYMPTOMS: Photodermatitis in about 48 hours after contact with the foliage with swelling and blisters; purple pigmentation may persist for a long time (171).

Mountain Laurel
Kalmia latifolia L.
Ericaceae

DESCRIPTION: A shrub or small tree, to 9 m tall.

LEAVES: Mostly alternate, or 2–3 together, simple, evergreen.

FLOWERS: Showy, many in branched clusters near the ends of branches; individual flower stalks long, sticky; calyx 5-lobed; petals 5, fused, white to pink; stamens 10; pistil 1; April–late June.

FRUIT: Small capsules with many small, light-brown seeds; Fall.

OCCURRENCE: An understory tree of eastern forests, also common on mountain slopes; sometimes cultivated.

TOXICITY: Andromedotoxin (72, 142) in all parts.

SYMPTOMS: Soon after ingestion: mouth, nose and eye watering; in 2–6 hours: nausea, vomiting, sweating, abdominal pain, headache, low blood pressure, slow pulse, drowsiness, weakness, tingling of the skin, incoordination, convulsions and increasing limb paralysis until death (72, 109, 142).

NOTES: Children have been poisoned sucking nectar from the flowers and drinking a tea made from the leaves (72, 109). Honey made from the nectar is poisonous but too bitter to be consumed in large amounts (72, 142). The plant has been used externally as an ointment for skin disorders and internally for neuralgic pains and as a sedative (113). The Delaware Indians used it for suicide (101, 109).

Rhododendron, Azalea

***Rhododendron* spp.**
Ericaceae

Rhododendron catawbiense

Rhododendron azalea

DESCRIPTION: Shrubs, rarely trees; with twisted stems and scaly bark.
LEAVES: Alternate, simple, deciduous or evergreen, entire or toothed, leathery or papery, often crowded near the ends of branchlets.
FLOWERS: Large, in loose to dense, terminal clusters; calyx small, 5-lobed or toothed, persisting in fruit; petals showy, widespread, trumpet to bell-shaped; stamens 5 or 10; pistil 1.
FRUIT: Elongated capsules, splitting into 5 sections at maturity; seeds very small, scale-like.
OCCURRENCE: Native and introduced; moist woods, borders, along streams; some cultivated.
TOXICITY: Grayanotoxin I and III (76) especially in the leaves; andromedotoxin (72, 142).
SYMPTOMS: See *Kalmia*.
NOTES: About 2.5 hours after ingesting an unknown amount of rhododendron leaves, a 4-year-old boy began to vomit and experienced abdominal pain. After becoming somnolent and less responsive, he was taken to a hospital for treatment which was successful (8).

Primrose
Primula obconica Hance
Primulaceae

DESCRIPTION: Perennial herb; stems scapose.
LEAVES: Basal, oblong-ovate, margins irregular, with glandular hairs.
FLOWERS: Regular, in umbels; sepals 5; corolla funnel-shaped, with 5 lobes; stamens 5; ovary 1-celled.
FRUIT: A capsule.
OCCURRENCE: Native to Eurasia; a winter blooming greenhouse plant.
TOXICITY: Glandular hairs on the leaves and stems contain the alkyl methoxy-quinone primin (55, 79, 109, 142).
SYMPTOMS: Some people experience allergic contact dermatitis when coming in contact with the plant (72, 106, 109, 142, 165).

Privet

Ligustrum vulgare L.
Oleaceae

DESCRIPTION: Tall shrubs up to 5 m tall.

LEAVES: Simple, glabrous, deciduous, 3–6 cm long.

FLOWERS: Small, white, in dense panicles; calyx and corolla tubular; stamens 2, inserted on the corolla tube, barely exserted or included; June.

FRUIT: Black berry, with 1–2 seeds.

OCCURRENCE: Native to Europe; naturalized in thickets and open woods, throughout the Northeast; sometimes cultivated.

TOXICITY: The glycosides ligustrin, ligustron, syringin, syringopicrin and other principles are found in the fruit (142, 165); foliage contains unknowns (142).

SYMPTOMS: Nausea, headache, abdominal pain, vomiting, diarrhea, weakness, irregular pulse, low blood pressure and body temperature with cold and clammy skin; severe cases show muscle spasms and convulsions; emanations from the flowers may cause respiratory irritation when in bloom (93, 128, 142).

NOTES: The berries persist throughout the winter when few others are present and thus may attract children. A decoction of the leaves or bark have been used for diarrhea and as a mouth or skin wash (113). In one case, a 5-year-old boy died several hours after eating the fruit (106).

Carolina Jessamine
Gelsemium sempervirens (L.) Aiton
Loganiaceae

DESCRIPTION: The state flower of South Carolina; a woody, evergreen, trailing or climbing vine, twining from left to right.
LEAVES: Opposite, simple, entire, lanceolate-elliptic, 1–5 cm long.
FLOWERS: Fragrant, solitary or in cymes, axillary; sepals 5, separate; corolla tubular, 5-lobed, yellow; stamens 5; pistil 1; March–early May.
FRUIT: Thin, flat, beaked capsule, to 2.5 cm long; September–November.
OCCURRENCE: Native in woods, thickets, fields of the southeastern Coastal Plain and Piedmont north to Virginia; sometimes cultivated.
TOXICITY: Alkaloids including sempervirine, gelsemine, gelseminine, gelsemicine, gelsemoidine and others occur throughout the plant (39, 72, 106, 108, 109, 128, 142, 163); the greatest concentrations occur in the roots and flower nectar (72).
SYMPTOMS: Sweating, weakness, pain above and in the eyes, dilated pupils, double vision, nausea, staggering, lowered temperature, convulsions, respiratory failure and death (49, 72, 101, 109, 128, 142); flowers, leaves and roots may cause dermatitis (72, 128, 165).
NOTES: Children have been poisoned chewing the leaves and sucking nectar from the flowers (49, 72, 101, 109). Rhizomes and root extracts have been used to reduce fever, relieve pain, as a central nervous system depressant and an emetic (102, 109), but overdoses have caused fatalities (142). Bees carrying the nectar will produce toxic honey (72). In one case of poisoning, a 3.5-year-old girl ate about 5 blossoms from a yellow jessamine vine growing on a fence in her yard at about 11:45 on an April morning. She was treated in an emergency room with syrup of ipecac, but when vomiting did not occur in 10 minutes, the yellow

petals were retrieved by lavaging the stomach. After this, a slurry of activated charcoal was given which she vomited at about 2:00 P.M. With continued treatment, she appeared well by about 9:15 that evening (23). Another alleged case involved the death of three persons in Branchville, South Carolina, in 1885, supposedly from consuming honey derived from Carolina jessamine. Medical journals, in reporting the case, stated that a large concentration of gelsemine was found in a sample of the honey used. However, a special inquiry by the USDA, Division of Botany, revealed that no chemical analysis had been done and that the diagnosis was based entirely upon symptoms. Therefore, the real cause of the case still remains an open question (35).

Indian Pink, Pinkroot

Spigelia marilandica L.
Loganiaceae

DESCRIPTION: Perennial herb with simple, erect stems; 3–6 dm tall.
LEAVES: Simple, opposite, entire, ovate-lanceolate, unstalked.
FLOWERS: In short, 1-sided spikes; sepals 5; corolla tubular, red outside, yellow inside, 5-lobed; stamens 5, exserted.
FRUIT: Two celled capsule; seeds few.
OCCURRENCE: Rich woods in Ohio and Maryland south to Florida, Missouri and Texas; sometimes cultivated.
TOXICITY: The alkaloid spigiline and other unknowns (129, 142).
SYMPTOMS: Increased circulation, dim vision, vomiting, dilated pupils, spasms of the eye and facial muscles and convulsions, sometimes followed by death (142).
NOTES: Serious reactions usually only occur when medicinal preparations are taken in excess (106). Indians used the plant to treat intestinal worms (113) as have people in Appalachia (102).

Yellow Allamanda
Allamanda cathartica L.
Apocynaceae

DESCRIPTION: Showy, fragrant shrubs.
LEAVES: Opposite or in whorls of 3–4, simple, elliptical, glossy, leathery, 10–16 cm long, with pinnate veins.
FLOWERS: In clusters at the ends of branches, yellow, aromatic, bell-shaped with 5 petal lobes, tubular.
FRUIT: Two-valved, prickly capsule; seeds winged.
OCCURRENCE: Native to tropical America; in disturbed areas of south Florida, sometimes cultivated.
TOXICITY: Fruit, stem and leaf sap contain unidentified toxins.
SYMPTOMS: Fever, swollen lips, thirst, dry mouth, nausea and purging (128); dermatitis on sensitive skin (72, 128, 142, 165).
NOTES: The cathartic effect is reportedly self-limiting (106, 142).

Madagascar Periwinkle
Catharanthus roseus (L.) G. Don
Apocynaceae

SYNONYM: *Vinca rosea* L.
DESCRIPTION: Erect herb; 2–6 dm tall.
LEAVES: Opposite, obovate-lanceolate, 3–7 cm long.
FLOWERS: One–2 in upper leaf axils; corolla tubular, pinkish-red, rarely white.
FRUIT: Sickle-shaped, pubescent follicles.
OCCURRENCE: Cultivated ornamental; seldom escaping.
TOXICITY: The indole alkaloids vinblastine, vincristine and others (97, 109, 142, 163).
SYMPTOMS: Smoking the dried leaves may cause incoordination, prickling of the skin and hallucinations (106); excessive or extended use may result in kidney and nervous system problems (142).
NOTES: The Latin name was *pervinca*, from *vincire*, which means to wind or entwine, in reference to the tendrils and rootlets. The name was modified in Middle English to periwinkle (48). The plant has been used to treat excessive menstruation, diarrhea and nose bleed (113). These alkaloids are also tumor chemotherapeutic (97, 163). Dose limiting side effects include leukopenia and neuro-muscular effects (163). The plant is also toxic to livestock (127).

Oleander

Nerium oleander L.
Apocynaceae

DESCRIPTION: Shrub or small tree to 10 m with thick, clear, gummy sap.
LEAVES: Opposite or whorled, entire, simple, leathery, narrow, veins light green and very noticeable.
FLOWERS: Clustered at the ends of branches; calyx tubular, 5-lobed; petals showy, 5, white to red or yellow; stamens 5; pistil 1.
FRUIT: Paired, leathery pods rarely forming; with small, winged seeds.
OCCURRENCE: Native to the Mediterranean cultivated in the East.
TOXICITY: All parts, especially the twigs, green or dry, leaves and flowers contain the cardiac glycosides neriin (nerioside) and oleandrin (oleandroside) (72, 109, 128, 142).
SYMPTOMS: After a few hours: dizziness, sleepiness, slow, irregular heartbeat, pupil dilation, nausea, vomiting, colic, bloody diarrhea, unconsciousness, convulsions, respiratory paralysis and death (72, 109, 128, 142); the leaves may also cause dermatitis (72, 90, 106, 109, 142, 165).
NOTES: This plant has been known as poisonous since ancient times (171). Drinking water from a vase that contained the flowers has caused poisoning (49). One leaf is said to be enough to kill an adult (72, 142), and poisonings have occurred when people ate hotdogs roasted on oleander stems over open fires (174). Breathing smoke from the burning foliage may also cause poisoning (128). Children have been poisoned by sucking nectar from the flowers and chewing the leaves (72, 109). Honey made from the nectar may also be poisonous (128, 142). The plant has been used in African arrow poisons as well (171). In one case, a previously healthy 96-year-old woman was found at home weak and vomiting. At an emergency room, her son gave a history of oleander leaf ingestion. Despite treatment, she died about 40 minutes later after eating 5 to 15 leaves (140).

Milkweed

***Asclepias* spp.**
Asclepiadaceae

Asclepias humistrata

Asclepias syriaca

Asclepias tuberosa

DESCRIPTION: Erect, perennial herbs with milky juice.

LEAVES: Opposite or whorled, rarely alternate, simple.

FLOWERS: In terminal or axillary umbels; sepals 5, reflexed, persisting; petals 5, white, yellow, greenish, red or purplish; stamens 5, filaments hooded and usually with incurved horns.

FRUIT: Inflated follicles with numerous seeds, each bearing a tuft of long silky hairs.

OCCURRENCE: Mostly native; throughout the East in woods, borders, old fields; some cultivated.

TOXICITY: Several cardiac glycosides of the 5-α series including uzarigenin and syriogenin (158) that are toxic to man (109).

SYMPTOMS: Some species have caused vomiting, stupor and weakness (142); the sap may cause dermatitis (109).

NOTES: The genus may be dangerous to children if eaten in quantity (72). The root of butterfly milkweed (*A. tuberosa* L.) is used medicinally, but overdoses are toxic (142). Nineteenth-century physicians used a tea from the powdered root as a mild sedative and for asthma (159). The root has also been used as a purgative, diuretic and emetic (113). Indians used milkweed for asthma, dropsy and stomach ailments. The juice was used externally for warts (159).

Palay Rubbervine
Cryptostegia grandiflora R. Br.
Asclepiadaceae

DESCRIPTION: Climbing shrub with milky sap.

LEAVES: Opposite, leathery, waxy, to 12.7 cm long.

FLOWERS: Funnel-shaped, pinkish-lavender, in terminal clusters.

FRUIT: Three-sided follicles, in pairs, to 10 cm long; seeds orange with a silky tip.

OCCURRENCE: Native to Madagascar, cultivated in south Florida, sometimes escaping.

TOXICITY: The cardiac glycosides cryptograndioside A & B occur in all parts, especially the milky sap (142).

SYMPTOMS: Small amounts internally have caused violent diarrhea, cardiac failure and death (142); the sap is also a skin irritant (135, 142, 165); dust from the dry vine is irritant to mucous membranes (128, 142).

NOTES: Both the Palay and Madagascar rubbervines (*C. madagascariensis* Bojer) were used to produce rubber in India in the 1800's. They were introduced into Florida as ornamentals, but research in the early 1930's found that mature leaves contain up to 3% rubber, which can be extracted by mechanical or chemical means. The latex has also been used by natives of Madagascar for murder and suicide. However, decoctions from the roots were used as a remedy for chronic blennorrhagias (144). Human deaths have also been recorded in India (109), and it has been used as an arrow poison in Africa (171).

Morning Glory
Ipomoea tricolor Cav.
Convulvulaceae

S. pseudocapsicum

SYNONYM: *I. violacea* L.
DESCRIPTION: Stout, twining, perennial vine.
LEAVES: Ovate-orbicular, to 25 cm across, cordate at the base.
FLOWERS: Corolla funnel-like, to 10 cm long; purplish-blue, tube white; cultivated varieties are variously colored.
FRUIT: A capsule with several seeds.
OCCURRENCE: Native to tropical America; cultivated varieties include "flying saucers" with dark blue and white striped flowers; "heavenly blues" with dark sky-blue flowers; and "pearly gates" with white flowers.
TOXICITY: Lysergic acid amide, isoergine, elymoclavine and other principles (43).
SYMPTOMS: Seeds cause hallucinations if taken in large quantities; side effects may include nausea, vomiting, diarrhea, drowsiness, numbness of extremities and muscle tightness (37).
NOTES: In one fatal case reported in 1964, a college student took several hundred seeds of the "heavenly blue" variety. He experienced an intense hallucinogenic experience which lasted about 24 hours. The symptoms recurred three weeks later, so he took a sedative in order to sleep. They also recurred several times during the next week. At this time, he committed suicide by crashing his car into a house at an estimated speed of 100 mph (37). The flowers, seeds, roots and stems of common morning glory (*I. purpurea* (L.) Roth) have been used as a laxative (102).

Lantana
Lantana camara L.
Verbenaceae

DESCRIPTION: Pubescent, sub-shrub; to 2.5 m tall; stems prickly, square.
LEAVES: Opposite or whorled, simple, toothed, ovate, 2–13 cm long, fragrant when crushed.
FLOWERS: In flat clusters on a long stalk; small, tubular, white, pink or yellow, changing to orange or red; May–frost.
FRUIT: Bluish-black, fleshy drupe.
OCCURRENCE: Native in dry woods of the Southeast; potted in the northern U.S. and Canada and planted outside in the southeastern Coastal Plain.
TOXICITY: The fruit contains lantanine (lantadene A and B), a hepatogenic photosensitizer in animals (91). It is most toxic when green (72, 128, 165).
SYMPTOMS: Within 2–5 hours in humans: lethargy, gastrointestinal upset with vomiting and diarrhea, dilated pupils, labored respiration, circulatory collapse and death (72, 101, 109, 128, 142); the leaves may cause dermatitis (128, 142).
NOTES: In Tampa, Florida, a previously healthy 3-year-old boy ingested an unknown amount of the green berries. He was taken to an emergency room 5 hours later with vomiting, diarrhea with the green berries in his stools, lethargy, cyanosis, labored breathing, dilated pupils and depressed deep tendon reflexes. He had gastric lavage and was treated with adrenal steroids, oxygen and croupette for 24 hours. Signs of toxicity lasted about 56 hours. He was discharged in 5 days with complete recovery. In an unrelated incident, his 2.5-year-old cousin ingested an

unknown amount of the berries. She was rushed to a hospital 6 hours later with similar symptoms, but died 90 minutes later of pulmonary edema and neuro-circulatory collapse. One single difference in the two cases is that the girl did not receive gastric lavage (177). All species are suspected of being toxic (72).

Belladonna, Deadly Nightshade

Atropa belladonna L.
Solanaceae

DESCRIPTION: Perennial herb; to 1.5 m tall.

LEAVES: Alternate, simple, entire, ovate, crowded on short branches.

FLOWERS: Purple, tubular, about 2.5 cm long; ovary superior; calyx 5-lobed, persisting in fruit.

FRUIT: Purplish-black berry.

OCCURRENCE: Native to Europe and Asia Minor; a garden ornamental in the U.S., sometimes naturalizing in waste areas of eastern states.

TOXICITY: Berries, leaves and roots contain tropane alkaloids (0.3–0.5%); mostly *l*-hyoscyamine (95–98%) with traces of *l*-scopolamine (hyosine) and astropine (*dl*-hyoscyamine); other toxins reported are apoatropine, belladonnine and cuscohygrine (108, 109).

SYMPTOMS: Fever, hot, dry, flushed skin, headache, dry mouth, thirst, difficulty in swallowing, burning of the throat, pupil dilation, visual disturbances, weak and rapid pulse, elevated blood pressure, urge but inability to urinate, constipation, delirium, hallucinations, incoordination, confusion, convulsions, coma with subnormal temperature, labored respiration, respiratory failure and death (72, 108, 109, 142); the fatal dose of atropine is estimated to be 10–20 mg in children and about 100 mg in adults, although adults have survived consumption of as much as 1 gm (10).

NOTES: Atropa has been known as a poisonous plant for thousands of years. One of the earliest references to it was in the Ebers Papyrus in 1550 B.C.E. (137). Belladonna is Italian for "beautiful lady," in reference to the juice being used at one time by women to enlarge the pupils of their eyes as a sign of beauty (108, 116). It was used with henbane

(*Hyoscyamus*) and jimsonweed (*Datura*) in the "flying ointments" of witches. Rubbing this concoction all over their bodies caused hallucinations of flying, etc. (116). In one case of poisoning 5 children were admitted to a hospital the day after a blackberry picking outing. Entwined among the prickly vines were several plants of belladonna with large black berries. During the night, they had delirium, visual difficulties, lost the use of their legs, were restless, talkative, had hot, dry skin, rapid pulse, dilated pupils and muscular incoordination. Three of the children were found to have eaten 20–40 berries, but none were removed from the other two who had equally severe symptoms. All five recovered from the incident (119).

Chili Pepper
Capsicum frutescens L.
Solanaceae

DESCRIPTION: Shrub up to 3 m tall.
LEAVES: Simple, variable in form and size, to 12 cm long.
FLOWERS: Yellow, lavender or white; petals 5, lobed.
FRUIT: Round or elongated, pointed pepper, changing from green to white, yellow, purple, finally red; to 5 cm long.
OCCURRENCE: Native and cultivated in tropical America; cultivated in the southern U.S. as an ornamental and for the fruit.
TOXICITY: Leaves contain solanine and other alkaloids (142); the fruit contains pungent principles, composed mostly of capsaicin and lesser amounts of hydrocapsaicin and nordihydrocapsaicin (109, 128).
SYMPTOMS: Even though used as condiments, the fresh fruits are very irritating to human tissues. A burning rash, redness, swelling and blistering may occur (72, 109, 128, 142, 165).
NOTES: Southwestern and Mexican herbalists have prescribed the peppers whole to cure colds and as a strong aphrodesiac (103). The smoke from burning peppers is used to irritate a victim's mucous membranes in one form of Malaysian torture (128).

Cestrum, Jessamine

***Cestrum* spp.**
Solanaceae

DESCRIPTION: Large, evergreen shrubs.

LEAVES: Alternate, simple, entire, lanceolate or elliptical.

FLOWERS: Trumpet-like, in axillary clusters; white, ivory or yellowish.

FRUIT: Small, white or purplish-black berry with several seeds.

OCCURRENCE: Night-blooming jessamine (*C. nocturnum* L.) and the fragrant day-blooming species (*C. diurnum* L.) are cultivated in the southern U.S.; also in thickets and waste areas, Georgia to Texas and south.

TOXICITY: Solanine type glyco-alkaloids are present in unripe berries and atropine-like alkaloids are more prevalent in mature berries (106, 142).

SYMPTOMS: Unripe berries cause nausea, irritated throat, abdominal pain, vomiting, bloody diarrhea, headache, salivation, fever, dilated pupils, sweating, stupor, difficult breathing, slow pulse, weakness, trembling, lowered temperature, convulsions, coma and death (142); ripe berries cause rapid heartbeat, hallucinations, incoordination, difficult respiration, fever, some paralysis, hot, dry skin and mouth, thirst, delirium, hallucinations, constipation and urinary retention (72, 142). Volatile emanations from the blooms may cause respiratory irritation (127, 128, 142).

Jimsonweed, Jamestown Weed
Datura stramonium L.
Solanaceae

DESCRIPTION: Annual herb, to 1.5 m tall.

LEAVES: Alternate, simple, ovate-elliptic, short-stalked, toothed, to 20 cm long.

FLOWERS: Showy, usually opening in the evening, solitary, terminal; calyx 3–5 cm long; Corolla tube-like, white or lavender, with 5 lobes; July–September.

FRUIT: Dry, many-seeded, spiny capsule with 4 locules.

OCCURRENCE: Native to Asia; naturalized in pastures, fields, waste areas, along roadsides; throughout North and South America; sometimes cultivated.

TOXICITY: All parts, including the pollen, but especially the leaves and seeds, contain the tropane alkaloids hyoscyamine, atropine and scopolamine in high concentrations (72, 97, 106, 109, 142).

SYMPTOMS: Hot, dry, and flushed skin, thirst, pupil dilation, headache, delirium, vomiting only rarely, incoordination and confusion, hallucinations, stupor, rapid and weak pulse, convulsions and coma with subnormal temperature; if death does not occur, severe symptoms usually subside after 12–48 hours, but pupil dilation may last 2 weeks (72, 97, 101, 142); fragrance from the flowers may cause respiratory irritation, headache, dizziness and nausea in some people (142); the sap may cause skin dermatitis (142).

NOTES: Four–5 grams have been reported as fatal to children who may suck nectar from the flowers, eat the seeds or make tea from the leaves (72, 101). In October 1955, a young Indiana farmer experienced pupil dilation and other visual difficulties with his left eye. The actual causative agent discovered by his wife was a jimsonweed seed in the corner of the eye (160). During Bacon's Rebellion, in Jamestown, Virginia, in 1676, soldiers sent to put down the uprising were poisoned by the fruit and cooked greens from this plant (72, 101). This gave rise to the colloquial name Jamestown weed with jimsonweed being a later slang corruption. In Europe, the seeds and plant extracts have been used to treat mania, epilepsy, rheumatism, convulsions and madness (109). People have been poisoned by its use in treating bronchitis and asthma; and in its use as a hallucinogen (101, 109). Many cases involving adults and adolescents have been described in recent years. In one case, a 15-year-old male was found nude with a flushed face and thorax; was incoherent and hallucinating. He was comatose and barely responsive to deep painful stimuli. The pupils were widely dilated. After recovery, he affirmed ingesting the seeds of jimsonweed (118).

Henbane, Black Henbane
Hyoscyamus niger L.
Solanaceae

DESCRIPTION: Erect, hairy annual or biennial; to 1 m tall.
LEAVES: Ovate-oblong, toothed; to 20 cm long.
FLOWERS: Bell-shaped, axillary; calyx persistent in fruit; corolla yellow, purple-veined; anthers purple; June–August.
FRUIT: Globular capsule.
OCCURRENCE: Native to Britain and Europe; grown as a medicinal herb in the U.S. and Canada; escaped in dry soils, disturbed waste areas and along roadsides; sometimes cultivated.
TOXICITY: All parts contain hyoscyamine, scopolamine and atropine (72, 101, 106, 109).
SYMPTOMS: Giddiness, nausea, headache, salivation, hallucinations, euphoria, increased pulse and blood pressure, dilated pupils, speech and vision disturbances, coma and death (4, 72); dermatitis (165).
NOTES: The seeds were used by the Babylonians 3000 years ago for toothache (163) and more recently in Europe (93). In the middle ages, it was mixed with ivy, poppy, hemlock and other herbs as an anesthetic for surgery patients (22). Henbane was also a main ingredient in love potions at one time. Hamlet's father was murdered by pouring a solution of henbane in his ear (116). Deaths have occurred from the roots being mistaken for chicory or parsnips and seeds eaten by children (134). Its toxicity may be distinguished from that caused by *Atropa* and *Datura* by the presence of excessive salivation (101).

Tomato

Lycopersicon esculentum Miller
Solanaceae

DESCRIPTION: Decumbent, strong-scented annual; stems freely branching.
LEAVES: Irregularly pinnately lobed, dentate; 1–5 dm long.
FLOWERS: In panicles; sepals 5, lanceolate; petals 5, yellow; June–frost.
FRUIT: Red or yellow, fleshy berry with many seeds.
OCCURRENCE: Native to South America; cultivated in gardens for the edible fruits; rarely escaping.
TOXICITY: The foliage and stems (not the ripe fruit) contain the solanidan alkaloids solanine and demissine and their aglycones solanidine and demissidine (97).
SYMPTOMS: Refer to *Solanum* for internal symptoms; allergic contact dermatitis (106, 142).
NOTES: Tomatoes were introduced to Europe in the 1500's, but were feared toxic because of their close relationship with nightshade and belladonna. Ketchup was finally accepted as a wholesome food in the U.S. about the middle of the 19th century (159). In Hawkins County, Tennessee, a family grafted tomato plants onto the roots of jimsonweed (*Datura stramonium* L.) to produce hardier tomatoes. The resulting fruit caused severe poisoning (101). In another case, a 45-year-old woman developed eruptions on her arms after picking tomatoes in her garden (105). Children have been poisoned by "decoctions" and "teas" made from the leaves (72, 109).

Tobacco
Nicotiana tabacum L.
Solanaceae

DESCRIPTION: Stout, erect annual; 3–30 dm tall; stems with sticky hairs.
LEAVES: Alternate, simple, entire, lanceolate, ovate or obovate, 1–6 dm long.
FLOWERS: In panicles; trumpet-like; petals 5, cream or white; July-frost.
FRUIT: Capsules with numerous minute seeds; September–October.
OCCURRENCE: Introduced; widely cultivated as a cash crop; a rare escape.
TOXICITY: All parts contain the alkaloid nicotine and others (109, 142, 163); N'–nitrosonornicotine, a potential organic carginogen has been isolated from unburned tobacco (81, 156).
SYMPTOMS: Muscular twitching, shaking, weakness, dizziness, abdominal pain, vomiting, diarrhea, slow or weak and rapid pulse, labored breathing, clammy skin, paralysis, respiratory failure and death (72, 109, 142); the fatal dose of nicotine in a 50 kg person is said to be about 40 mg orally (10, 96); the juice of the plant is a skin irritant (142, 165).
NOTES: Beside the controversial harmful effects of smoking, tobacco leaves are toxic as cooked greens, as a family found after eating wild tobacco (*N. trigonella* Dunal). In this case, the leaves were picked in a neighbor's yard. The mother and a 12-year-old daughter recovered after hospital treatment, but a 7-year-old girl died from nicotine poisoning (112). In most cases, the unavoidable tendency to vomit usually removes most of the material before great harm is done (10). Tobacco decoctions have caused poisoning when used in treating intestinal worms. In one case, a 2-year-old boy was given a decoction rectally. When attended by a physician, he was comatose, pulseless, had very faint breathing and dilated, insensitive pupils. Artificial respiration and soapy-water ene-

mas gradually brought about improvement. He vomited after 40 minutes and had recovered except for fatigue within 3 hours (56). In another case, a 2-year-old boy was given a tobacco enema by his grandmother, on the advice of a friend. Within 15 minutes, he began foaming at the mouth, vomited and had seizures. He then became limp and was taken to a hospital. Other symptoms displayed were dilated pupils, paleness, weak pulse, intercostal muscle paralysis with diaphragmatic breathing and unobtainable blood pressure. With treatment, he improved by the next morning (136). In a final case, a smuggler wrapped tobacco leaves around his body to avoid Customs duty on the product. However, perspiration moistened the leaves and caused him to absorb a lethal dose of nicotine through the skin (96, 171).

Trumpet Flower, Chalice Vine

Solandra spp.
Solanaceae

DESCRIPTION: Climbing shrub to 18 m tall.
LEAVES: Leathery, in rosettes.
FLOWERS: Ivory-white, turning yellow with purple streaks; fragrant at night.
FRUIT: Fleshy berry, with many seeds, white-yellow; edible.
OCCURRENCE: Cultivated in the southern U.S. as outside or houseplants.
TOXICITY: All parts except for the fruits contain the alkaloids solanine, solanidine and others that are atropine-like (106, 128, 142).
SYMPTOMS: Delayed gastroenteric irritation (106); refer to *Solanum* for details of solanine poisoning; ingestion of the flowers has caused incoordination, excitability, dilated pupils, swelling and numbness of hands and feet and delirium (128, 142); prolonged exposure to the fragrance during blooming may cause nausea and dizziness (128), the sap is an eye irritant (142).
NOTES: The flowers have been used as a hallucinogen (128), and children have been poisoned eating them and the leaves (72).

Nightshade
Solanum spp.
Solanaceae

S. pseudocapsicum

DESCRIPTION: Herbs, vines or shrubs; covered with small star-shaped hairs.
LEAVES: Alternate, simple, entire to lobed or parted.
FLOWERS: Perfect; white, yellow, blue or purple.
FRUIT: Berries with dry or juicy pulp and several seeds.
OCCURRENCE: Native and introduced; waste areas, borders, pastures, roadsides; some cultivated.
TOXICITY: The solanidan alkaloids solanine and demissine (97); solanine occurs in all parts but is most concentrated in the immature fruit. Upon hydrolysis brought about by being bruised or eaten, it yields the sugar solanose and the alkamine aglycone solanidine (72).
SYMPTOMS: Intact, the alkalod solanine is not readily absorbed from the gastrointestinal tract (49), but is irritant and causes nausea, vomiting, abdominal pain and constipation or diarrhea. Its aglycone solanidine causes nervous effects such as apathy, drowsiness, salivation, weakness or paralysis, circulatory and respiratory depression, unconsciousness and death (109).
NOTES: Bittersweet (*S. dulcamara* L.) root bark and twigs have been used in a poultice for herpes, and black nightshade leaves (*S. nigrum* L.) leaves have been used internally for purging and to promote perspiration (113). Rappahannock Indians steeped a few leaves of common nightshade (*S. americanum* Miller) in water as a cure for insomnia (109). Carolina horse nettle (*S. carolinense* L.) was linked to the death of a 6-year-old boy in Delaware County, Pennsylvania, in January 1963 (101). Jerusalem cherry (*S. pseudocapsicum* L.) which is sold as a potted

Nightshade

S. tuberosum

S. carolinense

plant for its lush foliage and bright red berries has caused human poisoning (142). Common potatoes (*S. tuberosum* L.) that have spoiled or turned green, as well as the sprouts, should be avoided since they have caused severe poisoning. A case occurred on October 18, 1924, when a family began using 1.5 bushels of potatoes that had turned green after being exposed to sunlight. Eight members of the family began showing signs of poisoning two days later. The mother, aged 45, died on October 25, and a daughter, aged 16, died on the 27th (70).

Foxglove

Digitalis purpurea L.
Scrophulariaceae

DESCRIPTION: Erect biennial.

LEAVES: Alternate, simple, toothed.

FLOWERS: In terminal racemes, tubular, pendent, purple, pink, red, white or yellow, with spots inside the corolla tube.

FRUIT: Dry capsule.

OCCURRENCE: Native to Europe, a garden ornamental and medicinal plant in eastern America.

TOXICITY: Leaves, seeds and flowers contain cardiac glycosides including digitalin, digitoxin, gitalin, digitophyllin, digitalein, digitonin, digoxin and gitaloxin (109, 110, 142); drying does not diminish the potency of the toxins (93).

SYMPTOMS: Nausea, vomiting, diarrhea, abdominal pain, headache, drowsiness, fatigue, irregular heartbeat and pulse, tremors, convulsions and death (72, 101, 109, 142).

NOTES: The effects of foxglove were first noticed in England in the 1700's when William Withering, a Shropshire physician, saw cures brought about by an aged herbwoman (48, 93). Even today, this is still the most effective drug for congestive heart failure (10). Poisoning in adults usually results from overdoses of the drug, while children are poisoned by sucking the flowers for nectar or chewing the leaves or seeds (72, 109). Drinking water from a vase containing the cut flowers has also caused poisoning (106). The slightly sweet taste of glycosides further increases the possibility of ingestion by children (10). The plant has been used as an arrow poison in Africa (171).

Trumpet Creeper, Cow-Itch Vine

Campsis radicans (L.) Seemann
Bignoniaceae

DESCRIPTION: Woody climbing vine.
LEAVES: Opposite, pinnately divided; leaflets toothed.
FLOWERS: In clusters; trumpet-like, yellowish-orange to red; 5-lobed, 5–7.5 cm long; June–September.
FRUIT: Long, slender capsule with many, winged seeds.
OCCURRENCE: Native to the eastern U.S.; in thickets, moist or dry woods, on fences and trees; may be cultivated.
TOXICITY: Unknown.
SYMPTOMS: Contact with the leaves or flowers may result in skin irritation with blisters (72, 109, 138, 142, 143, 165); various parts may be toxic or irritating if ingested (142).

Japanese Honeysuckle

Lonicera japonica Thunb.
Caprifoliaceae

DESCRIPTION: Climbing or trailing vine, glabrous to pubescent.

LEAVES: Opposite, entire, evergreen.

FLOWERS: Fragrant, paired, subtended by 2 bracts; calyx 5-lobed; corolla 2-lipped, white or pink-yellow; stamens 5; April–July.

FRUIT: Black, glossy berries, with flattened elliptic seeds.

OCCURRENCE: Introduced; sometimes cultivated; now escaped and sometimes problematic by strangling and covering native vegetation; throughout the East in thickets, borders, pastures, on fences, etc.

TOXICITY: Known chemical components: vine with saponin, tannin, ash and HCN; flowers with luteolin and i-inositol (126); ripe fruit with 8 carotenoids, mainly cryptoxanthin and γ-carotene (171).

SYMPTOMS: Soon after ingestion of the berries: severe emesis, colic, diarrhea, pupil dilation, cold sweat, accelerated heartbeat; twitching of the limbs may be followed by convulsions, respiratory failure, coma and death; recovery in non-fatal cases is usually 3 or more days (106).

NOTES: The leaves have been used as a tea substitute in China and as a diuretic in Vietnam (171). The dried flowers have been used as a diuretic in Japan (126). Children frequently suck the nectar of the flowers with no apparent adverse effects.

Elderberry
Sambucus spp.
Caprifoliaceae

DESCRIPTION: Woody plants, with stout branches and a soft pith.
LEAVES: Large, opposite, deciduous to almost evergreen, pinnately compound.
FLOWERS: Small, perfect, white, yellow or pinkish in large, flat-topped clusters; calyx 3–5 lobed; petals 3–5, fused near the base; stamens 5; pistil 1.
FRUIT: Globose, juicy berry with several seeds.
OCCURRENCE: Moist woods and margins; widely distributed; some cultivated.
TOXICITY: Toxic alkaloids and cyanogenic glycosides that release HCN occur in the leaves, stems, bark and roots (72, 109, 142).
SYMPTOMS: Gastroenteritis, nausea, vomiting and diarrhea (49, 72); refer to *Prunus* for details of cyanide poisoning.
NOTES: Children have been poisoned by eating the unripe or uncooked berries or roots (109) and by playing with peashooters from the hollowed out stems (49, 72, 109, 142). Iroquois Indians boiled the inner bark of *S. canadensis* L. to treat toothache, ulcers, burns and so forth (5). Pioneers also used the bark in an ointment for chafed skin, abrasions and burns (102). An infusion of the fresh flowers has been used to wash sore eyes; steeped dried flowers have been used for fever and the leaves as a cathartic (5). Fresh ripe berries are used safely in pies, pancakes, jellies and wine (72, 142).

Balsam Pear, Bitter Gourd

Momordica charantia L.
Cucurbitaceae

DESCRIPTION: Climbing, annual vine with a foul smell; stems grooved, somewhat hairy, 6–9 m long.

LEAVES: Palmately lobed, to 15 cm across.

FLOWERS: Unisexual, yellow, tubular.

FRUIT: Yellowish-orange, warty; pulp red; seeds up to about 20, thick, with a red fleshy aril.

OCCURRENCE: Native to Old World tropics; naturalized in waste areas, sandy soils of the Southeast; sometimes cultivated as a porch vine.

TOXICITY: The outer fruit coat and seeds contain a resin, a saponic glycoside yielding elaterin and toxic alkaloids (109). *M. charantia* L. var. *abbreviata* Ser. seeds contain the bitter principle momordicin; ripe fruits contain a hypoglycemic agent charantin (128).

SYMPTOMS: Small amounts cause headache, facial redness, salivation, pupil dilation, nausea, vomiting, diarrhea and abdominal pain; serious cases may show muscular weakness and death (49, 109, 142).

NOTES: Infusions from the ripe fruit and vine have been used to induce abortion (109, 142). The almost ripe fruit has been pickled whole in whisky with rock sugar and used as a cough syrup in Missouri and the Ozark Plateau (109).

Cardinal Flower, Indian Tobacco
Lobelia spp.
Lobeliaceae

L. cardinalis   *L. siphilitica*

DESCRIPTION: Erect annual or perennial herbs.

LEAVES: Alternate, simple, entire or toothed.

FLOWERS: Axillary, in terminal racemes, variously colored; petals tubular, irregular, with 2 lobes forming an upper lip, 3 forming the lower one.

FRUIT: Capsules with many seeds.

OCCURRENCE: Native in old fields, margins, open woods, throughout the U.S.; may be cultivated.

TOXICITY: All parts contain the alkaloids lobelanine and lobeline (72, 106, 109, 142, 163); lobelidine is in all above ground parts (108); other toxins include the volatile oil lobelinan (109), nicotine and other pyridine alkaloids (97).

SYMPTOMS: Nausea, vomiting, weakness, tremors, sweating, rapid and weak pulse, headache, depressed temperature, stupor, collapse, convulsions, coma, paralysis and death (72, 101, 109, 142).

NOTES: Indians used the leaves and flowering tops of *L. inflata* L. as a fast acting emetic and purgative (109). *L. cardinalis* L. and *L. siphilitica* L. were used in the 19th century to promote sweating and to break up and help discharge bronchial congestion (109). Poisonings usually result from overdoses of such medicines (72, 101, 109, 142). Externally, the plant has been used in a poultice for bruises, sprains, ringworm, insect bites and poison ivy rash (113). Lobeline salts are used as the active ingredient in several types of smoking deterrents (108, 163).

Milfoil, Yarrow

Achillea millefolium L.
Asteraceae

DESCRIPTION: Erect, perennial herb with rhizomes; 3–12 dm tall.
LEAVES: Alternate, 2–3x pinnately divided, with many basal rosettes.
FLOWERS: Composite heads in terminal clusters; each head subtended by several series of involucral bracts; ray flowers pistillate, white or pink; disc flowers perfect, white, rarely pink; pappus none; late April–September.
FRUIT: Achenes.
OCCURRENCE: Meadows, pastures, waste places; throughout the U.S.; sometimes cultivated.
TOXICITY: Unidentified.
SYMPTOMS: Allergic contact dermatitis in some people (106, 108); it may also cause percutaneous photosensitization (171).
NOTES: The plant was named in honor of Achilles, who in legend used it in healing his soldiers wounds (159). Medically, it has been used for appetite loss, stomach, liver and gallbladder ailments and in a wash for skin disorders such as sores and chapped hands (113). Blackfoot Indians used *Achillea* leaves and flowers in a decoction as an eyewash; Winnebago Indians steeped the plant and poured the liquid into aching ears (109). The root was used by Indians as a local anesthetic in cleaning wounds and for toothache (5). If eaten by cows, the plant also causes a bad flavor in milk (129).

White Snakeroot
Ageratina altissima (L.)
King & Rob Asteraceae

SYNONYM: *Eupatorium rugosum* Houtt.
DESCRIPTION: Perennial herb; to 1.2 m tall.
LEAVES: Opposite, simple ovate, coarsely toothed.
FLOWERS: Small, white, in clustered, composite heads; receptacle flat; pappus of bristles; late July–October.
FRUIT: Achenes.
OCCURRENCE: Roadsides, margins, open woods, fields; Canada south to Georgia and east Texas.
TOXICITY: All parts contain trematol, a complex, unstable alcohol (142), in combination with a resin (72, 109) and glycosides (72).
SYMPTOMS: After a day or so of weakness: nausea, tremors, abdominal pain, vomiting, delirium, constipation, acetone odor on the breath, collapse, coma and death (72, 109, 142).
NOTES: In colonial times, this plant caused an illness known as *milk sickness,* and was brought on by drinking milk from cows that had eaten it (72, 109, 142). The cause of the illness was unknown during the time of the greatest number of deaths in the early 1800's (101, 142). It was most prevalent in dry years when there was a shortage of proper forage (101). Poisonings usually resulted from the daily consumption of small amounts of contaminated milk (142). It has been suggested as the cause of death of Abraham Lincoln's mother (101). Current animal husbandry practices and pooling of milk from many producers ensures prevention of the disease today, but poisonings are still possible where milk is used from a family cow.

Sneezeweed
Helenium spp.
Asteraceae

H. autumnale

H. amarum

DESCRIPTION: Erect annual or perennial herbs; stems often branched.

LEAVES: Alternate, cauline and basal, sometimes decurrent, often with aromatic, resinous globules.

FLOWERS: Many, in heads; ray flowers mostly fertile, some pistillate, 3–5 lobed at the apex; disc flowers perfect, yellow or reddish-purple; involucral bracts small, reflexed.

FRUIT: Achenes.

OCCURRENCE: Some native on banks, in ditches, pastures, margins, northern Michigan and south; some are cultivated.

TOXICITY: The allergenic sesquiterpene lactone tenulin has been found in *H. amarum* (Raf.) H. Rock (88); helenalin has been found in *H. autumnale* L. (55).

SYMPTOMS: Noted in animals: weakness, trembling, salivation, rapid and irregular pulse, labored breathing, vomiting, diarrhea, spasms, convulsions and death; liver and kidney damage have been noted in autopsies of livestock (142). All species may cause dermatitis (109, 165).

NOTES: Wheat contaminated with large amounts of the seeds has caused human poisoning when consumed in baking products (90, 101, 142). The plant has been used to induce sneezing to clear nasal passages of mucous and also as a fish poison (102). It also causes a bitter flavor in cow's milk.

Senecio, Squaw Weed
Senecio spp.
Asteraceae

DESCRIPTION: Erect, annual or perennial herbs.
LEAVES: Basal or cauline, alternate, lobed or unlobed, some pinnate.
FLOWERS: In solitary or corymbose heads; involucre cylindric to bell-shaped; disc flowers yellow, perfect; ray flowers pistillate, 3–5 toothed, or none present.
FRUIT: Achenes.
OCCURRENCE: Native and introduced: swamps, roadsides, margins, overgrazed pastures; some cultivated.
TOXICITY: Pyrrolizidine alkaloids are found in different species. Senecionine, seneciophylline, jacobine, jaconine, jacoline and jacozine have been found in tansy ragwort (*S. jacobea* L.) of the Northeast and West (41).
SYMPTOMS: From extended use as a medicinal tea: loss of appetite, vomiting, bloody diarrhea, sleepiness, weakness, staggering and jaundice; serious cases result in liver damage and death (142); some species may cause dermatitis (142).
NOTES: Chronic consumption of foods contaminated with pyrrolizidine alkaloid containing species as well as herbal medicines have been a problem in South Africa. These cases have shown acute veno-occlusive lesions which progress to cirrhosis of the liver (41). Such use in the West Indies has caused abdominal pain, diarrhea, enlarged liver, bloody vomiting, dropsy and death in 2 weeks to 2 years time (128). Honey made with nectar from tansy ragwort has been found to contain pyrrolizidine alkaloids. Although toxic, it is very bitter and large amounts would probably not be consumed. However, long term consumption of

palatable honey with low concentrations may be harmful (41, 45). These alkaloids may also be passed in the milk of dairy animals and thus cause problems where a family cow is used (42). An herbal tea known as "gordolobo yerba" from *S. douglasii* DC (synonym: *S. longibus* Benth.) has caused poisoning in the southwestern U.S. In one case, a 6-month-old Mexican-American female was admitted to a hospital with vomiting, hepatomegaly, ascites and right plural effusion. It was found that the child had been given 8 oz. of the tea on alternate days for 2 weeks prior to admission. In another case, a 62-year-old Mexican-American died of complications of hypertension, cirrhosis, portal hypertension and hepatic encephalopathy. She had consumed several cups of gordolobo tea a day for 6 months (86). Mexican flame vine (*S. confusus* Britten) of south Florida, has irritant sap, which is a source of dermatitis when being trimmed (72, 127, 128, 142, 165).

Tansy

Tanacetum vulgare L.
Asteraceae

DESCRIPTION: Erect, strong-smelling, perennial herbs with rhizomes; to 9.5 dm tall.

LEAVES: Alternate, simple, pinnately divided with narrow, toothed segments.

FLOWERS: In flat, composite heads; yellow; ray flowers none; disc flowers small; pappus none; July–October.

FRUIT: Achenes.

OCCURRENCE: Cultivated in herb gardens; escaped along roadsides, in fields, and waste areas; throughout North America.

TOXICITY: Tansy oil contains thujone and other principles (106, 109).

SYMPTOMS: Rapid but weak pulse, severe gastritis, convulsions and death (72, 101, 129); the sap may cause dermatitis (106, 109, 165).

NOTES: *Tanacetum* is from the French *tanasie*, which came from the Greek word *althanasia*, meaning immortality (48). Legend has it that a drink made from the plant was used to make the young Ganymede "immortal," so that he could serve as Zeus' cup-bearer (113). A tea from the leaves and flowers called "oil of tansy" was once used for abortions, intestinal worms and to promote menstruation (72). It has been described as worse than the ailments it was intended to cure (48) and overdoses have caused death (72, 113). Externally, it has been used in a wash for skin disorders and sprains (113). The plant was once hung in colonial kitchens and dried as a pepper substitute and used to help keep insects away (5).

MISCELLANEOUS

(Additional species for which only limited information and/or case histories could be found.)

Adonis vernalis L. Spring adonis (Ranunculaceae) is a perennial herb with alternate leaves that is cultivated for the large showy flowers. Rhizomes and roots contain the glycosides cymarin, adonitoxin and *k*-strophanthin (109). Toxic symptoms include nervousness, stomach upset, depression and even death in large amounts (72). This species has been used as an arrow poison in Africa (171).

Aethusa cynapium L. Fool's parsley (Apiaceae) is an annual herb with hollow stems, pinnate leaves and small white flowers in umbels that was native to Europe and now occurs on waste and cultivated grounds of the Northeast. All parts are toxic with aethusin, aethusanol, cynapine and traces of coniine (4, 93). Symptoms include burning of the mouth and throat, nausea, abdominal pain, vomiting, diarrhea and dizziness; serious cases may lead to unconsciousness, coma, convulsions, circulatory and/or respiratory failure and death (93). Poisonings have occurred due to the leaves being mistaken for parsley and the roots for turnips or radishes (134). In one case of poisoning, a woman added the leaves to soup. Two children who ate it died within 24 hours (93).

Anacardium occidentale L. Cashew nut is from a tree in the cashew family (Anacardiaceae) which also has poison ivy in it. The fruit shell contains urushiol, cardol and anacardic acid, which will blister the skin on contact (109, 128, 142, 171). The toxin is removed in a heating process before the nuts are freed (128), though some people still get mouth and lip irritation when handling and eating them. Fumes from the roasting shells are irritating to the face and eyes (142). It is cultivated in Florida, California and Arizona. The bark of the tree has been used in African arrow poisons and oil from the fresh nut has been used in Tanganyika for making tribal markings on the skin (171).

Apium graveolens L. Celery (Apiaceae) is a perennial commerical crop that is a rare escape from cultivation in eastern America. The seed oil contains about 60% *d*-limonene and 10% selinene (108). Photodermatitis has been reported (106, 109). In a study of farm workers near Boston who handled celery, exposed skin areas first itched, and when scratched lesions appeared and then blisters in 12–24 hours. The blisters later ruptured, dried, cracked and lichenified. Dark-skinned laborers from Jamaica and the Bahamas seemed immune and about ⅓ of the Caucasians were affected. The oil of the plant was shown to cause the irritation (176).

Arnica montana L. Leopard's bane (Asteraceae) or European arnica is a hairy, erect perennial herb with yellow flowers in composite heads, that is native to Europe and cultivated in gardens of the northern U.S. and Canada. The flowers and roots with unidentified toxins, cause mucous

Miscellaneous

membrane irritation, diarrhea, vomiting, giddiness, weakness, increase or decrease in pulse rate and collapse. Allergic contact dermatitis may also occur (108). An extract from the plant has been used as an irritant (109), but one oz. (30 ml) of the tincture may cause serious symptoms (108).

Baptisia tinctoria (L.) R. Br.   False indigo (Leguminoseae) is a perennial, much-branched herb with smooth, blue-gray leaves and pea-like flowers. It is native in dry woods and open areas throughout the East. Human poisoning has occurred from taking overdoses of tincture of Baptisia (113, 129). The active principles are baptisin and cytisine (106, 142). Plant extracts have been used externally for wounds and skin disorders (113).

Bryonia alba L.   White bryony (Cucurbitaceae) is a prickly perennial, climbing plant that is cultivated in the U.S. It has rough, 5-lobed leaves and white-greenish flowers that produce black berries. The spindle-shaped rootstock which contains a milky sap, is used as a purgative. When dried, it has been for ailments such as whooping cough. In large doses, the rootstock is poisonous. Forty berries reportedly cause adult fatalities and 15 in children (113).

Caesalpinia gilliesii (Hook.)   Wallich ex Benth. synonym: *Poinciana gilliesii* Hook.   Bird-of-paradise (Leguminoseae) or poinciana is a shrub or small tree up to 5 m with alternate, bipinnate leaves and light yellow flowers in terminal racemes. The flower stalks are covered with sticky hairs. The fruit is a legume that splits explosively at maturity to scatter the 5–7 seeds. It is native to South America and cultivated in the extreme southern U.S. but has escaped along roadsides and in disturbed areas from Texas to California. In September 1957, two boys in Tucson, Arizona, each ingested about 5 seed pods. Within ½ hour, both had nausea, vomiting and profuse diarrhea which recurred intermittently for 24 hours (7). Gardeners should therefore remove the pods to avoid accidental poisonings.

Calotropis procera (Aiton) Dryand in W.T. Ait.   Giant milkweed (Asclepiadaceae) is an erect perennial with opposite, blue-green leaves and pink or white flowers that produce follicles. It is common in the tropics and cultivated in Florida. The milky latex contains calotoxin, calactin, calotropin, uscharidin, uscharin, vovuscharin and calotropogenin (74), gomphroside (27), proceroside (29) and others. Small amounts have been used externally for skin problems such has leprosy in India (109), but it may irritate sensitive areas (142). The toxins have also been used in African arrow poisons (171). Ingestion of large amounts may cause vomiting, diarrhea, labored breathing, elevated blood pressure, slow but stronger heartbeat, convulsions and death (142).

Celastrus scandens L.   Climbing bittersweet (Celastraceae) is a deciduous, twining, climbing vine up to 18 m long, with alternate, simple, serrated leaves, and unisexual flowers. The orange capsules have seeds that are covered by a pulpy red-orange aril. It occurs on moist soil along roads and streams in woods from Canada south to North Carolina, and is also cultivated. The roots, fruits and leaves cause chills, weakness, vomiting, diarrhea, convulsions and coma (90). The root bark has been used to induce vomiting and treat venereal disease (102). Dried arrangements with the plant should be kept out of the reach of children (109).

Coriaria spp. Coriaria (Coriaceae) is a genus of deciduous shrubs and perennial herbs with opposite, entire leaves. The 5-parted flowers give rise to red, yellow or black berry-like fruits with 5, one-seeded nutlets. It is native to New Zealand, Asia and South America and now cultivated in the U.S. The fruit of some species contains the convulsant coriamyrtin (39) and may cause rapid heartbeat, labored breathing and convulsions (106). The fruit of *C. myrtifolia* L. has caused death in children (106). In the Phillipines, a family of four reportedly died after taking a decoction from one of the species (154).

Daphne mezereum L. Daphne (Thymelaceae) is a deciduous shrub up to 12 dm tall with alternate, simple leaves and fragrant, lilac-purple flowers in small clusters that give rise to long, leathery, scarlet drupes. It is cultivated and sometimes escaped in the Northeast. All parts contain mezerein, a daphnane ester (55, 104, 105). Symptoms include internal irritation with swelling of the lips and tongue, thirst, salivation, difficulty in swallowing, nausea, vomiting, abdominal pain, bloody diarrhea and weakness; severe cases may show delirium, collapse, convulsions, coma and death (4, 72, 101, 109, 134). Blisters may arise if the leaves are rubbed against the skin (90, 101, 106, 165). It is thought that 10–12 fruits will be fatal to a child (4). This plant has been considered toxic for centuries. An alternative to the old practice of bleeding with leeches was to place small patches of the bark soaked in vinegar on the skin to cause blistering and thus draw out unwanted "humors" (93). It has also been used to treat cancers for centuries. Recent research showed an alcohol-water extract from the plant to be anti-leukemic in certain cases in mice (104).

Dioscorea bulbifera L. Air potato (Dioscoreaceae) is an aggressive, high climbing, annual vine that is grown ornamentally in south Florida. It has ovate leaves, but flowers are usually not produced here. The underground tubers contain diosgenin a steroidal sapogenin, the alkaloid dioscorine and a norditerpene lactone known as diosbulbine (94, 128). They are poisonous when eaten raw, though edible when properly prepared (128, 142). Toxicity includes gastroenteritis, nausea and bloody diarrhea (142). The aerial tubers are toxic when immature (128).

Duranta repens L. Golden drop (Verbenaceae) is a tree or shrub with ovate leaves to 5 cm long and 5-parted, white-purple flowers in loose clusters that give rise to yellow, globular, juicy, few-seeded berries. It is native to tropical America including Key West, but is cultivated in south Florida and grown in greenhouses elsewhere. Saponins in the fruit and foliage (90, 101, 109, 127) cause gastroenteric irritation, chiefly to the intestinal mucosa, drowsiness, fever, nausea, vomiting, convulsions and even death in serious cases (72, 106, 142); dermatitis may also occur if

Miscellaneous

handled (142). Children have been poisoned by the fruits with some deaths (72, 101, 109).

Equisetum arvense L.   Horsetail (Equisetaceae) is a small perennial herb of moist places throughout the U.S. with hollow, jointed stems. All parts are used medicinally, but overdoses may be toxic (113). Do not allow children to play with whistles or other toys made from the stems.

Ervatamia coronaria Stapf. synonym: *Tabernaemontana divaricata* (L.) R.Br.   Crape jessamine (Apocynaceae) is a shrub native to India that is cultivated in Florida. It has waxy flowers that are very fragrant. The leaves, stems, bark and milky juice have unidentified toxins, but the roots contain the alkaloids tabernaemontanine and coronarine (142) and are considered poisonous (128). The roots and leaves have been used medicinally (127).

Erythrina herbacea L.   Eastern coral bean (Leguminoseae) is a perennial herb with prickly branchlets that is found on sandy soils in open woods and clearings of the southeast Coastal Plain. Seeds contain the alkaloids erysodine, erysovine, erysopine, erysothiopine, erysothiovine and hypaphorine (128, 142) and have been used to poison rats and dogs in Mexico (128). The seeds of coral tree (*E. variegata* L. *var. orientalis* Merr.) of south Florida contain erythraline and hypaphorine (128). The leaves and bark have been used medicinally, but they contain a narcotic alkaloid erythrinine that causes CNS depression. HCN is found in the fruits, leaves, stems and roots. The raw seeds have poisoned children with diarrhea and vomiting (142). All species are potentially poisonous to children (109).

Fagus spp.   The American beech (*F. grandiflora* Ehrh) (Fagaceae) is reported as edible, but the raw nuts should not be eaten in quantity for beechnuts have caused poisoning in Europe (72, 101, 109). The seeds contain a saponic glycoside (109) and the pollen of the cultivated beech (*F. sylvatica* L.) is said to cause hay fever (90). Large amounts of the seeds may cause gastrointestinal upset (101, 109), but wildlife readily eat them when mature (19). Pioneers used a decoction of the fresh or dried leaves of American beech for burns, scalds and frostbite (102). Indians ate the small nuts for kidney ailments (5).

Gossypium spp. (cotton) (Malvaceae)   In the late 1950's, medical researchers in eastern China noticed a high incidence of childless marriages among people consuming large quantities of crude cottonseed oil from cooking. Tests begun on rats in 1971 confirmed that gossypol, a yellowish disesquiterpene aldehyde found in the seed, stem and root of the cotton plant is spermicidal. Soon after this, workers in Jiangsu Province began testing it as a possible oral contraceptive for men. Over 10,000 volunteers were tested between 1972–1978, and it proved 99.89% effective with minimal side effects. Subjects were given a daily dose of 20 mg and most became infertile in 50–70 days (173). After discontinuation of gossypol, it usually took about 3 months for sperm count and morphology to return to normal. A few transient side effects noticed in the early days of drug administration were general weakness, an increase or decrease in appetite, epigastric discomfort, nausea, and a decrease in sexual drive. Potency, however, was not impaired. A few individuals experienced potassium deficiency (36). At higher doses (100–700 times the amount required to prevent conception), the agent has been shown to cause hair discoloration, diarrhea, malnutrition, circulatory problems and even heart failure (115).

Invitro tests to determine if gossypol can be used as a vaginal contraceptive concluded it did not decrease sperm motility. This was also the case with gossypol acetic acid. However, its water soluble co-precipitate polyvinylpyrrolidone resulted in complete immobilization of sperm in varying lengths of time depending on concentration (170).

Guaicum officinale L. Lignum vitae (Zygophyllaceae) is a small evergreen tree native to the American tropics that is occasionally cultivated in south Florida as an ornamental. It yields a valuable wood that is hard and heavy. The resin guaicum is used as a chemical indicator, in medicines and stains (113, 142). If consumed in quantity, it is toxic (109, 142). The leaves are said to be acrid (142).

Heliotropium spp. Annual or perennial herbs (Boraginaceae) with alternate leaves and flowers in terminal racemes. They occur in various habitats throughout the East. All parts contain pyrrolizidine alkaloids such as heliotrine and dehydroheliotrine (97). *H. europaeum* L. contains heliotrine and lasiocarpine (109). Its use in herbal teas for extended periods may cause liver damage and other complications (109). This genus has also caused problems as a contaminant of grain. A serious epidemic involving 1600 recorded cases of veno-occlusive disease occurred in Afghanistan in 1976, when bread was made from wheat contaminated with *Heliotropium* and *Crotolaria*. Many of these people died. A similar outbreak that year in India involved 68 people with 28 deaths (86).

Heracleum mantegazzianum Somm. & Levier Giant hogweed or cow parsnip (Apiaceae) is a biennial or perennial herb with hollow stems that are usually blotched; alternate, compound leaves up to 9 dm long, and white or pink flowers in umbels. It is native to the Caucasus area between the Black and Caspian Seas and was introduced into the U.S. as an ornamental. It has naturalized in waste areas, along roads and streambanks; occurring in at least 12 countries of central and western New York; found recently near Montpelier, Vermont. Furocoumarins in the sap (55, 106) cause severe photodermatitis after contaminated skin is exposed to the sun (106, 165). Resulting scars may last for years (72). Several cases of dermatitis due to this plant were reported in Edinburgh, Scotland. In June 1970, nine children, aged 11–14, used hollow stems from the plant growing in a park as peashooters and telescopes. About 36 hours later, all had developed redness and itching of exposed areas and some developed blisters in the next 12 hours. Medical examination at 72 hours after exposure revealed large blisters with clear fluid. After treatment, healing was rapid in all cases (47). Since this species is a

Miscellaneous

public health hazard and is of very limited distribution in the U.S., it has been classified as a federal noxious weed under the Federal Noxious Weed Act of 1974. This prohibits further importation of the species into the United States.

Illicium floridanum Ellis   A small evergreen tree (Illiaceae) native to damp ravines and swamp borders of northwest Florida, south Alabama, Mississippi and Louisiana. The leaves are toxic and cause gastroenteritis, vomiting and convulsions (128). Japanese anise (*I. anisatum* L.) which is native to the Far East is cultivated in the southeastern U.S. It has a toxin similar to cicutoxin in water hemlock that causes convulsions, respiratory and circulatory failure and death (128). The seeds contain star anise oil which has been used medicinally (113). The principal constituent is safrole (4-allyl, 2 methylene dioxybenzene) (100) which has caused liver cancer in rats (82).

Ipomoea alba L. [synonym: *Calonyction aculeatum* (L.) House] (Convulvulaceae—moonflower, moonvine) is an abundant, wild vine in Florida. It is used to cover fences because of its night-blooming flowers. Handling or trimming the prickly stems which have milky sap may cause dermatitis in some people (128).

Lablab niger Medik (synonym: *Dolichos lablab* L.)   Hyacinth bean (Leguminoseae) is an annual, twining, climbing vine with alternate, trifoliate leaflets and racemose white, pink or purple flowers that produce flat legumes with black or white seeds. Native in the Old World tropics, it is widely cultivated in South America for food and has been introduced in the U.S. When cooked, the water should be changed several times, for the raw beans contain cyanogenic glycosides that release HCN (109, 142). Eaten raw, they cause weakness, vomiting, labored breathing, twitching, stupor, convulsions and unconsciousness (142).

Lolium temulentum L.   Darnel (Gramineae) or tares, is an annual grass of grain fields and waste areas that is rarely cultivated.

Toxic principles include the alkaloids temuline, loline and temulentic acid (142, 171). Poisoning symptoms may include apathy, giddiness, weakness, drowsiness, trembling, headache, pupil dilation, delirium, dizziness, confusion, nausea, vomiting, coma, respiratory paralysis and death (101, 129, 142, 171). Darnel has been considered toxic for thousands of years. Human poisoning is suspected from flour contaminated with the seeds (101, 171). It has been suggested that the plant is made toxic only when infected by the saprophytic fungus *Endoconidium temulentum* Prill and Del. (101, 171). It has been used medicinally for stomach disorders and skin ailments (113).

Manihot esculenta Crantz (synonym: *M. ultissima* Pohl)   Cassava (Euphorbiaceae) is a bushy herb or shrub with elongated tubers and deeply lobed leaves. The panicles flowers give rise to 6-angled, globose capsules. It is native to Brazil, but widely cultivated for the tubers that serve as the basic carbohydrate source for 300 million people in the tropics and alone meet 8–10% of the daily global caloric needs (97). Tapioca is made from them. The tubers are cyanogenic unless properly prepared (72, 165), with as much as 150 mg of HCN per kg of fresh weight (38). Leaves and tubers contain the cyanogenic glycosides linamarin and lotaustralin (97). See *Prunus* for details of cyanide poisoning. The plant is naturalized in disturbed areas of south Florida and cultivated as well.

Manilkara zapota (L.) V. Royen   Sapodilla (Sapotaceae) is a flowering plant native to southern Mexico that is cultivated in south Florida. The milky latex is nontoxic (108) and the fruit pulp is edible when ripe (142). The bark and seeds contain the glycoside sapotin that causes gastroenteritis (142).

Mentha pulegium L.   European pennyroyal (Lamiaceae) grows throughout the East. It was used in ancient Rome to induce abortions and has recently found use as a herbal tea to induce menses and for its calming and diaphoretic effects. One case involved an 18-year-old woman who had been using herbal preparations for two years. Two hours after ingesting 30 ml of the oil, she was taken to a hospital in a coma. She died a week later of liver and kidney failure (68).

Mucuna deeringiana (Bort) Merrill (synonym: *Stizolobium deeringianum* Bort)   Velvet bean (Leguminoseae) is a twining vine with compound leaves and greenish-yellow flowers that produce dark brown seed pods. It is native to Asia and naturalized in old fields of south Florida. Even after extended boiling (which gives off noxious, irritating odors), the seeds taste bad and cause nausea and digestive upset (142). Cowhage [*M. pruriens* (L.) DC] is a similar species with stinging hairs that penetrate the skin causing itching and irritation (55).

Myristica fragrans Houtt.   Nutmeg (Myristicaceae) is the dried seed of the nutmeg tree; mace is the dried, orange-yellow aril that surrounds the shell which encloses the seed. It has been used as a spice for at least 2000 years, with definite evidence of its appearance in Europe at the end of the 12th century. Medicinally, it has been used to combat pain, vomiting and intestinal disorders, among other things (172). The small amounts used in cooking are harmless, but sufficient doses cause hallucinogenic effects and larger amounts result in nausea, vomiting, dryness of the mouth, stupor, disorientation, flushing, rapid heartbeat and even death (108). The essential oil contains myristicin, safrole a carcinogen and other principles (100, 172).

Pedilanthus tithymaloides (L.) Poit.   Redbird-cactus (Euphorbiaceae) or slipper flower is a succulent shrub with milky juice, alternate leaves and red flowers in clusters that give rise to small capsular fruits. It is native to tropical America, but is cultivated in warm climates and occurs in hammocks and pinelands of south Florida. All parts eaten in quantity will cause vomiting, diarrhea and gastroenteric irritation (106, 109, 128, 142). Externally it may cause dermatitis and blistering of the skin (72, 142). The plant has been used medicinally in Mexico and Japan (135), and as an emetic in the West Indies (128).

Miscellaneous

Phaseolus lunatus L. (Java Bean—Leguminoseae) These plump, colored, beans can be distinguished from common lima or navy beans by distinct lines radiating from the scar at the center of the inner curved side. It is native to the tropics but now cultivated in the southern U.S. as an ornamental. Some varieties of this bean contain the cyanogenic glycoside linamarin and/or phaseolunatin which form HCN in dangerous concentrations (72, 109, 142, 169). The roots are reported toxic as well (72). HCN content of the small black lima bean of Central America has been found to be as much as 400 mg/100 g of tissue and has caused human poisoning (122). U.S. varieties of lima or butter bean generally contain less than 0.01% cyanide, a level considered safe for consumption (145). Problems may rise from consumption of ornamental varieties normally cultivated for the flowers and foliage. Cooking reduces the toxin, but may not completely eliminate it because of the beans' thickness and denseness (109). Kidney beans have also been shown to contain small amounts of cyanogen (2.0 mg/100 g of tissue) (145).

Plumbago indica L. Plumbago (Plumbaginaceae) is a flowering shrub cultivated in south Florida with trumpet-shaped flowers in spike-like panicles. The leaves, stems and roots contain plumbagin and oil of plumbago. The toxins are very irritant and cause blisters on the skin (166). The root has been used as an emetic (142).

Pongamia pinnata (L.) Merrill Pongam (Leguminoseae) is a woody climber or tree with smooth gray bark and yellowish lenticels (corky spots on the bark, corresponding to epidermal stomata). It is native to India but now cultivated in south Florida. The leaves are pinnate and the flowers produce short, woody, 1-seeded pods. The seeds and roots which contain pongamiin and other principles are used in India to poison fish (72, 127, 142). The seed oil and other parts are used medicinally (127, 142). Toxic symptoms are gastroenteric irritation chiefly of the intestinal mucosa (106).

Pteridium aquilinum (L.) Kuhn Bracken (polypodiaceae) and extracts from it have carcinogenic properties. The toxins may be passed into milk and could present a health hazard to people in marginal farming areas of the world where cattle are known to consume it (54). The carcinogen has also been found in young leaves after being cooked for consumption (53). The plant has been used to treat intestinal worms and diarrhea. Indians used it as a diuretic and to relieve stomach cramps (102).

Portulaca oleracea L. Purslane (Portulacaceae) and wood sorrel (*Oxalis spp.*—Oxalidaceae) may contain high levels of oxalates (109, 142). See rhubarb (*Rheum*) for details of toxicity. If used as potherbs, a small amount of baking soda may be added and the cooking water changed to help neutralize the acid (142).

Raphanus sativus L. Cultivated radish (Brassicaceae) has been known to cause acute contact dermatitis. The roots and seeds yields mustard oils (isothiocyanates) which are used as a condiment and rubefacient. In one case, a 38-year-old waitress developed acute dermatitis after 3 weeks of chopping salad plants in a restaurant where she worked. Positive patch tests were obtained to the type of radishes she handled at work. The irritant chemicals were found to be allyl isothiocyanate and benzyl isothiocyanate (120). Ingestion of wild radish (*R. raphanistrum* L.) may cause gastrointestinal irritation, abdominal pain and bloody diarrhea. The leaves should be thoroughly cooked before eaten as a green (142).

Rhodotypos scandens (Thunb.) Makino (synonym: *R. tetrapetala* Makino) Jetberry bush (Rosaceae) is a deciduous shrub native to Japan and China that is cultivated in the U.S., occasionally escaping. It has opposite, serrated leaves and solitary flowers that give rise to large black achenes in the fall and winter. The fruits contain amygdalin and other cyanogenic glycosides that release HCN (109). Refer to *Prunus* for details of cyanide poisoning. On the evening of August 25, 1952, a 5-year-old girl began vomiting and had abdominal pains. The next day she was admitted to a hospital in a semi-comatose state with a fever of 105° F and dilated but reactive pupils. The glucose level which was found to be 290 mg/100 cc and a CO_2 combining power of 14 volume percent suggested a serious infection and diabetic coma. Insulin therapy, however, was withheld because a urinalysis prior to hospitalization was normal. She was given distilled water and 1/6 molar lactate intravenously as well as penicillin, terramycin and adrenal cortical extract. At 8:20 P.M., her temperature rose to 106° F and she began convulsing; which was controlled by phenobarbitol. An hour later, the glucose level had fallen to 210 mg/100 cc without insulin therapy. The temperature returned to normal within 48 hours of admission and she was discharged on the 5th day. A week later she confessed to having eaten fruits from a bush in a park just before becoming ill. Samples of the plant and fruit were identified as jetberry (147).

Thevetia peruviana (Pers.) K. Schum   Yellow oleander (Apocynaceae) or be-still tree is a shrub or tree up to 10 m tall with alternate, glossy, dark green leaves and tubular, yellow-orange flowers that give rise to fleshy, triangular, yellow-orange drupes. It is native to tropical America and now cultivated in the southern U.S. All parts except the fruit pulp contain the cardiac glycosides thevatin A and B, neriifolin, peruvoside, thevatoxin and others (9, 34, 101, 109, 127, 128, 135, 142). Poisoning symptoms include numbness and burning of the mouth, throat dryness, vomiting, dilated pupils, slow irregular heartbeat, high blood pressure, convulsions, cardiac arrest and death. The sap may also cause skin and eye irritation (142). Most fatalities are due to the plant's misuse in folk medicine (109, 142). One-2 seeds have been fatal to children (109). It has also been used as an arrow poison in Africa (171).

Thuja occidentalis L.   American arborvitae (Cupressaceae) is an evergreen tree with small scale-like leaves that occurs from Canada to North Carolina and is cultivated. Oil from the leaf contains up to 65% thujone and other principles, including isothujone, borneol, and pinene. Ingested in quantity, it causes hypotension (abnormally low blood pressure), con-

Miscellaneous

vulsions and death. It has been used internally as an diuretic and emmenagogue; externally for skin diseases and as an insect repellent (108).

Thymus vulgaris L.   Garden thyme (Lamiaceae) is a small plant widely cultivated as a spice. Medicinally, it has been used for throat and bronchial problems, diarrhea, colic and other ailments. Excessive use of thyme oil is toxic and may over-stimulate the thyroid gland (113). Other symptoms include nausea, vomiting, gastric pain, headache, dizziness, convulsions, coma, cardiac and respiratory collapse (108); it may also irritate the skin (109).

Vitex trifoliata L. f.   An aromatic hedge plant (Verbenaceae) in south Florida. The leaves contain cineol, 1-d-pinene, camphene, terpinyl acetate and a diterpene alcohol (128). Some people experience respiratory irritation, dizziness, headache and nausea when trimming it (142). The leaves are sometimes burned as an insect repellent (142).

READY REFERENCE LIST

A search of current literature yielded this list of vascular plants of Eastern America known or suspected to cause human poisoning. Literature citations are given for verification and easy reference when needed. Information is coded as follows:

    C: cultivated.
    N: native or naturalized.
    P: Photodermatitis.
    AA: airborne induced allergic reactions such as hay fever from pollen or respiratory distress caused by volatile emanations from blossoms or foliage.
    I: internal poisoning.
    D: dermatitis.
    #'s: literature citations.

Different authors' information on each plant is coded as follows:

(examples)
    4: poisoning only.
    4s: poisoning suspected.
    4d: poisoning with deaths recorded.
    4*: animal poisoning.
    4?: not specific about human poisoning.

Abelmoschus esculentus (L.) Moench. (okra) C; irritant hairs on most varieties. D: 142, 165.

Abies balsamea (L.) Mill. (balsam fir) CN; D: 106, 109.

Abrus precatorius L. (precatory bean) CN; see text p. 77.

Acacia spp. N; notes: thorn scratches may cause dermatitis (108); pollen may cause hay fever (109); dust from the tree may cause skin lesions and severe asthma in susceptible people; it is nontoxic ingested orally (108).

Acer spp. (maple) CN; oleoresins in pollen and sap may cause contact dermatitis (106, 109); AA: 109.

A. negundo L. (box elder) CN; oleoresins (109); AA: 106; D: 106, 109, 142.

Achillea millefolium L. (yarrow) CN; see text p. 167.

Aconitum spp. CN; see text p. 54.

A. napellus L. (garden monkshood) see text p. 54.

A. reclinatum Gray CN; I: 72.

A. uncinatum L. CN; I: 72.

Acorus calamus L. CN; oil of calamus; one major constituent is β-asarone which has caused intestinal tumors in rats at high doses (65, 100); may also contain crystals of calcium oxalate since it is in the arum family (Araceae).

Actea spp. (baneberry) CN; see text p. 55.

A. pachypoda Elliott (white baneberry) CN; all parts but mostly roots and berries; I: 72, 138d; D: 106.

A. rubra (Ait.) Willd. (red baneberry) as above.

Adenanthera pavonina L. (beadtree, red sandalwood) CN; raw seeds with lignoceric acid; I: 142s.

Ready Reference List

Adonis aestivalis L. CN; all parts; I: 72, 138.

A. vernalis L. CN; see p. 173.

Aesculus spp. (buckeye, horse chestnut) CN; see text p. 115.

A. flava Soland. ex Hope (yellow buckeye) synonym: *A. octandra* Marsh; CN; all parts; I: 72.

A. glabra Willd. CN; leaves, flowers, sprouts, seeds I: 72, 138.

A. hippocastanum Nutt. CN; gastroenteric irritant, chiefly to intestinal mucosa (106); I: 106, 138.

A. octandra Marsh CN; see *Aesculus flava.*

A. parviflora Walt. (bottlebrush buckeye) CN; I: 142s.

A. pavia L. (red buckeye); CN; all parts, especially the seeds and leaves; I: 49d, 126, 138, 142sd.

A. sylvatica Bartr. (painted buckeye) CN; all parts especially the seeds and leaves; I: 72.

Aethusa cynapium L. (fool's parsley) N; see p. 173.

Agarista populifolia (Lam.) Judd (fetter bush) synonym: *Leucothoe populifolia* (Lam.) Dipel N; all parts; see *Kalmia* p. 133; I: 142s.

Agave spp. (century plant) CN; see text p. 33.

A. picta Salm.-Dyck C; herb used medicinally; eating leaves or roots may cause fatality (113); D: 128.

A. sisalana Perr. (sisal agave) C; sap; D: 127.

Ageratina altissima (L.) King & Rob (white snakeroot) synonym: *Eupatorium rugosum* Houtt.; N; see text p. 168.

Aglaonema commutatum Schott C; D: 128.

A. modestum Schott ex Engl. (Chinese evergreen) C; D: 128.

Agrimonia spp. (agrimony) C; D: 106.

A. eupatorium L. C; irritant sap; P: 109; D: 106, 165.

Agropyron repens (L.) Beauvois (quack grass) N; pollen, AA: 109; D: 106.

Agrostemma githago L. (corn cockle) CN; see text p. 52.

Agrostis spp. (red top) N; pollen, AA: 109.

Ailanthus altissima (Miller) Swingle (tree of heaven) CN; leaves and flowers, I: 138, 142; D: 72, 142, 165; pollen, AA: 109; mechanical irritation (165).

Aleurites cordata (Thunb.) R. Br. (Japan wood-oil tree) C; all parts, especially seeds with a toxalbumin and saponin, I, D: 142s.

A. fordii Hemsl. (tung oil tree) CN; see text p. 96.

A. moluccana (L.) Willd. (candlenut) CN; all parts especially seeds; I: 106, 128, 138, 142; D: 106s, 142; with a toxalbumin and saponin (142).

A. montana (Lour.) Wils. (mu-oil tree) CN; all parts with a toxalbumin and saponin; I, D: 142.

Alisma plantago-aquatica L. (mad dog weed) CN; tubers; allergic contact dermatitis (165).

Allamanda spp. CN; may cause irritation and allergic reactions as well (109).

A. cathartica L. CN; see text p. 140.

A. violacea Gardner & Field CN; all parts, especially sap, I, D: 142.

Allium canadense L. (meadow leek); CN; all parts, I: 90?.

A. cepa L. (wild onion) CN; allergic contact dermatitis (106, 165); the mild irritant properties may be due to propenyl sulphenic acid (55).

A. cernuum Roth CN; all parts; I: 90?.

A. sativum L. (garlic) C; irritant juice may cause allergic contact dermatitis (106, 109, 165).

Alnus spp. (alder) CN; pollen; AA: 109.

Alocasia spp. (giant elephant ear) C; all parts with calcium oxalate crystals; gastroenteric irritant chiefly of mouth and throat; I: 106, 109, 128, 138, 165; D: 106, 128.

A. macorrhiza (L.) G. Don C; all parts with calcium oxalate crystals, especially corms; I: 127, 128, 138, 142; D: 127, 128, 142, 165; the cyanogen triglochin has been reported (84).

Amaranthus spp. (pigweed) N; pollen; AA: 109.

Amaryllis spp. C; see text p. 34.

Amaryllis belladonna L. (belladonna lily) C; bulbs and seeds with lycorine (142); I: 90, 138, 142s, 165.

Ambrosia spp. (ragweed) N; oleoresins and lactones; AA: 90, 109; I: 106; D: 106, 109.

A. ambrosioides (Cav.) Payne N; allergic contact dermatitis; sesquiterpene lactone damsin (55).

A. artemisifolia L. N; pollen, leaves and stems with allergic sesquiterpene lactones (55); AA: 142, 165; D: 72, 142, 165.

A. bidentata Michx. (lance-leaved ragweed) N; D: 106.

A. ludoviciana Nutt. N; allergic contact dermatitis from sesquiterpene lactones douglanine, ludovicin A, ludovicin B, ludovicin C (55).

Amianthium muscaetoxicum (Walter) Gray synonym: *Zigadenus muscaetoxicum* (Walt.) Zimm. CN; all parts with zygadenine, zygacine and others (142); I: 138, 142, 143.

Ammi majus L. (greater ammi, bishop's weed) CN; furocoumarins; P: 109, 165.

Ampelopsis arborea (L.) Koehn (peppervine) CN; leaves and berries with unknowns; I: 138; D: 142s.

Anacardium occidentale L. (cashew nut) C; see p. 173.

Anagallis arvensis L. (scarlet pimpernel) CN; fruit husk may cause allergic contact dermatitis (165); leaves with saponin and cyclamin, a glycoside (142); plant hairs with primin (142); I: 90, 113, 138, 143; D: 72, 113, 138, 142, 165.

Ananas comosus (L.) Merrill (pineapple) C; sap with bromelin (142) sometimes causes mild dermatitis among harvesters (106); D: 72, 113, 138, 142, 165.

Andira inermis (Wright) HBK ex DC. (cabbage bark tree) C; bark and seeds with andirine and other alkaloids; medicinal overdoses may be fatal (142).

Andromeda spp. CN; fruit with andromedotoxin (165); I: 138, 165; see *Kalmia* p. 133.

Anemone spp. (windflower) CN; see text p. 56.

A. caroliniana Walt. (pasque flower) CN; all parts with protoanemonin; I, D: 142.

Anethum graveolens L. (dill) C; limonene, a monocyclic terpene; D: 109.

Angadenia berterii (A. DC.) Miers (pineland allamanda) N; milky sap in all parts with unknowns; D: 142.

Angelica spp. CN; AA: 109.

Anisostichus capreolata (L.) Bureau synonym: *Bignonia capreolata* L. (crossvine) CN; all parts with unknowns; I: 142s.

Annona spp. C; seeds, I: 142s; irritant juice, D: 109.

A. cherimola Mill. (sugar apple) C; seeds, I: 142s.

A. glabra L. (pond apple) C; seeds, I: 142s.

A. muricata L. (sour sop) C; seeds, I: 142s.

A. reticulata L. (custard apple) C; seeds, I: 142s.

A. squamosa L. (sugar apple) C; seeds, I: 142s, 127?.

Anthemis spp. CN; lactones, D: 109.

A. arvensis L. (mayweed) N; leaves, flowers, D: 142s; all parts, D: 165.

A. cotula L. N; leaves and flowers with an acrid substance (142); I: 72; D: 72, 106, 142, 165.

Anthriscus spp. (chervil) CN; pollen, AA: 109s; allergic reactions upon contact with the flowers, I: 109.

Ready Reference List

A. sylvestris (L.) Hoffm. (wild chervil) CN; furocoumarins; P, D: 106, 165.

Anthurium spp. (tailflower) C; all parts with calcium oxalate crystals (142); I: 109, 142s; D: 142s.

Antirrhinum spp. (snapdragon) C; leaves, I: 138, 165.

Apium graveolens L. (celery) C; see p. 173.

Apocynum spp. (dogbane) CN; may be irritating and cause allergic reactions as well (109); D: 109.

A. androsaemifolium L. (spreading dogbane) CN; rootstock medicinal; CAUTION: has killed animals (113*).

A. cannabium L. (Indian hemp) CN; all parts toxic; glycosides apocannoside (apocynein), apocynin and cymarin (109, 142); I: 72s, 90, 126*, 138, 142s.

Aporocactus flagelliformis (L.) Lam. (rattail cactus) C; ornamental species with irritant sap (128).

Aquilegia vulgaris L. (European columbine) CN; seeds; has been fatal (109); I: 109d, 134, 138, 165.

Aralia spp. CN; irritant hairs on some species, D: 109.

A. spinosa L. (Hercules club) CN; see text p. 125.

Arctium spp. CN; D: 109.

Arecastrum romanzoffianum (Cham.) Becc. (queen palm) C; unripe seeds, I: 127?.

Argemone albiflora Hornemann (white prickly poppy) N; seeds, leaves and roots with alkaloids berberine, protopine; I: 142s.

A. mexicana L. (Mexican prickle poppy) CN; see text p. 68.

Arisaema spp. CN; D: 109.

A. dracontium (L.) Schott (dragon root) CN; all parts with calcium oxalate crystals (106, 142); I: 106, 142; D: 106, 142s.

A. triphyllum (L.) Schott (jack-in-the-pulpit) CN; see text p. 11.

Aristolochia clematis L. (birthwort) N; all parts medicinal, overdoses toxic; toxin acts similar to colchicine; I: 113.

A. serpentaria L. (Virginia snakeroot) CN; roots medicinal; alkaloid aristolochicine (142), toxic in pure form (113); bitter principles aristolochin, serpentarine (142); I: 113, 142s.

Armoracia rusticana Gaertn., Meyer & Scherb. (horseradish) C; sap with sinigrine; D: 165.

Arnica montana L. (European arnica) CN; see p. 173.

Artemisia spp. (white sage) CN; pollen, AA: 109; lactones, D: 109; I: 138, 165sd.

A. absinthium L. (wormseed) CN; flower and leaves, D: 109, 165; overdoses of the medicinal oil are toxic (113); thujone (109); I: 113, 138.

A. ludoviciana Nuttall (Louisiana wormwood) CN; D: 106.

A. vulgaris L. (mugwort) CN; rootstock, herb used medicinally; overdoses are toxic (113); I: 113; D: 106.

Asarum canadense L. (wild ginger) CN; leaves; D: 72, 165.

Asclepias spp. (milkweed) CN; see text p. 143.

A. curassavica L. (scarlet milkweed) CN; all parts with asclepiadin (142); uzarin, calactin, calotropin and other cardenolides (162); I: 135*, 142*; D: 142.

A. tuberosa L. (butterfly milkweed) CN; all parts with glycosides, resins; used medicinally; I: 142, 143.

Asimina spp. (pawpaw) CN; fruit, I: 142s.

A. parviflora (Michx.) Dunal N; fruit, I: 142s.

A. triloba (L.) Dunal CN; see text p. 63.

Asparagus officinalis L. C; see text p. 22.

A. sprengeri Regel (asparagus fern) C; foliage, fruit and seeds, I: 142s.

Ready Reference List

Aster spp. CN; pollen, AA: 109; allergic contact dermatitis (106, 109).

Atriplex spp. (beach oracle, saltbush) CN; all parts with saponins and soluble oxalates, I: 142s.

Atropa belladonna L. (belladonna, deadly nightshade) CN; see text p. 148.

Avicenna germinans (L.) L. (black mangrove) N; seeds toxic (127, 142s), edible if cooked (127).

Avena spp. (oats) C; pollen, AA: 109.

Bambusa vulgaris Schrad. ex Wendl. (common bamboo) CN; irritant bristles, D: 109.

Baptisia spp. (false indigo) CN; medicinal overdoses are toxic, I: 142s.

B. tinctoria (L.) R. Br. (wild indigo) CN; see p. 174.

Belamcanda chinensis (L.) DC. (blackberry lily) C; all parts with unknowns, I: 142s.

Berberis spp. CN; I: 138?; D: 165s.

Berchemia scandens (Hill) K. Koch (supple-jack) CN; all parts, especially berries; I: 142s.

Bertholettia excelsa Humb. & Bonpl. (Brazil nut) C; nuts during processing; seed oil; D: 106, 142.

Beta spp. (sugar beet) C; pollen may cause bronchial asthma (109).

B. vulgaris L. (beet, Swiss chard) C; leaves with soluble oxalates; large amounts toxic (142).

Betula spp. (birch) CN; pollen, AA: 109.

Bignonia capreolata L. CN; see *Anisostichus*.

Blighia sapida Koenig (akee) C; see text p. 117.

Borago officinalis L. (borage) CN; herbs, flowers medicinal; fresh leaves, D: 113.

Brassica spp. CN; irritant oils; P, D: 109.

B. nigra (L.) W. Koch (black mustard) C; sap with sinigrine (165) is toxic in large amounts; mustard oil with isothiocyanates (55) is a powerful irritant; may cause blisters and/or lacrimation (108); I: 90; D: 165.

B. rapa L. (turnip) C; all parts in some people, D: 165.

Broussonetia spp. (paper mulberry) N; pollen, AA: 109.

Brugsmansia x candida Pers. synonym: *Datura candida* (Pasq.) Saff. C; all parts with atropine, scopolamine and hyoscyamine (106, 127, 142); I: 106, 127, 128, 138, 142d; D: 142.

B. suaveolens Humb. & Bonpl. ex Willd. synonym: *Datura suaveolens* Humb. & Bonpl. ex Willd. (angel's trumpet) C; all parts; I: 90, 106, 127, 128, 142d; D: 142.

Bryonia alba L. (white bryony) C; see p. 174.

Buxus sp. (boxwood) C; D: 109.

B. microphylla Sieb. & Zucc. (Japanese boxwood) C; all parts with buxine; I: 142s.

B. sempervirens L. (common boxwood) C; see text p. 107.

Caesalpinia bonduc (L.) Roxb. (wait-a-bit vine); N; all parts medicinal esp. seeds; bonducin; prickly; I: 142.

C. gilliesii (Hook.) Wallich ex Benth. (bird-of-paradise) C; see p. 174.

C. pulcherrima (L.) Swartz (Barbados pride, flower fence) C; leaves: resin, gallic acid, HCN; leaves, fruit, flowers with tannin; all parts medicinal; I, D: 142.

Caladium spp. C; see text p. 12.

C. bicolor (Ait.) Vent. C; all parts, esp. corms with calcium oxalate (109, 142); I: 44*s, 128, 142; D: 128, 142.

C. picturatum Koch & Bouche C; calcium oxalate, I, D: 142.

Calla palustris L. (wild calla, water arum) CN; calcium oxalate crystals (106); gastroenteric irritant, chiefly mouth and throat (106); I: 30, 106, 109, 143; D: 30, 106.

Calonyction aculeatum (L.) House CN; see *Ipomoea* on p. 178.

Calophyllum brasiliensis Camb. synonym: *C. calaba* Jacq. (Brazil beauty leaf) C; I: 142s.

C. calaba Jacq. C; see *Calophyllum brasiliensis*.

C. inophyllum L. (Indian laurel) C; see text p. 121.

Calotropis spp. C; D: 109.

C. gigantea (L.) Ait. f. (crownflower) C; milky juice with calotropin, calactin, calotoxin, uscharidin, uscharin (127, 142, 165); all parts medicinal; I: 72s, 127, 138d, 142d, 165; D: 106, 127, 138, 142, 165.

C. procera (Ait.) Dryand in Aiton C; see p. 174.

Caltha palustris L. (cowslip, marsh marigold) N; see text p. 57.

Calycanthus spp. (Carolina allspice) CN; seeds convulsant; with calycanthidine, calycanthine; I: 106.

Campsis radians (L.) Seemann ex Bur. (trumpet creeper, cow-itch) CN; see text p. 162.

Cannabis sativa L. (hemp, marijuana) CN; see text p. 42.

Capsella bursa-pastoris (L.) Medik. (shepherd's purse) N; seeds; D: 165.

Capsicum anuum L. var. *aviculare* (Dierb.) D'Arcy & Eshb. CN; leaves with solanine and others; used medicinally and as a condiment (142); I: 142s; D: 142s, 165.

C. frutesciens L. (chili pepper) CN; see text p. 150.

Cardiospermum halicacabum L. (balloon vine) CN; sap; D: 165.

Carex spp. (sedge) CN; pollen; AA: 109.

C. arenaria L. (sea sedge) N; rootstock medicinal; should not be used if kidney inflammation is present (113).

Carica papaya L. (papaya) CN; irritant sap with proteolytic enzyme papain and chymopapain (55, 142); leaves with carpain, carposide (142); seeds with carpasemine, carcin, myrosin (142); may cause allergic reactions in some people; widely used in tenderizing meat (108); pollen, AA: 109; D: 72, 109, 142, 165.

Carpinus spp. (muscle tree) CN; pollen; AA: 109.

Carum carvi L. (caraway) C; limonene; AA, D: 109.

Carya spp. (hickory) CN; pollen; AA: 109.

Caryota spp. (fishtail palm) C; I: 165d.

C. mitis Lour. C; see text p. 10.

C. urens L. C; fruit juice and pulp; fibers at base may be irritating; I: 142s; D: 127, 128, 142s.

Casimiroa edulis Llave (white sapote) C; seeds are narcotic and toxic; I: 127, 142s.

C. tetrameria Millsp. (matasano) C; seeds are narcotic and toxic; I: 142s.

Cassia spp. (senna) CN; leaves and other parts with anthraquinones (142) and cyanogenic glycosides that release HCN (109); I: 109, 142.

C. alata L. (candlebush, ringworm cassia) CN; leaves with an anthraquinone; fruit with oxymethylanthraquinones; all parts with some HCN; leaves and fruits medicinal; I: 142s.

C. bicapsularis L. (Christmas senna) CN; anthraquinones, purgative; toxic in large amounts; I: 142s.

C. fasciculata Michx. (partridge pea) CN; leaves and seeds with anthraquinones; purgative; may be toxic in large amounts; I: 142s.

C. fistula L. (golden shower) C; fruit pulp, leaves used as purgative; leaves with anthraquinones; fruit pulp with emodin glycosides, oxymethylanthraquinones (106, 142); gastroenteric irritant chiefly to intestinal mucosa (106); I: 138, 142.

C. marilandica L. (wild senna) CN; large amounts of leaves with anthraquinones used as purgative may be toxic; I: 142s.

C. obtusifolia L. (sickle pod) N; large amounts of leaves with anthraquinones used as laxative and for skin diseases may be toxic; I: 142s.

C. occidentalis L. (coffee senna) CN; see text p. 78.

C. siamea Lam. (kassod tree) C; leaves, pods, bark, roots, I: 142, 135*; pockets of yellow powder in bark are irritant to skin and eyes, with chrysophanhydroenthron (128, 142).

Cassytha filiformis L. (love vine) C; laurotetanine (142); all parts used medicinally in Africa and India, large amounts toxic (142); I: 142sd, 165d.

Casuarina spp. (Australian pine) CN; AA: 109.

Catalpa spp. CN; flowers, D: 72.

C. bignonioides Walter (bean tree, catawba tree) CN; flowers with unknown irritant, D: 142s.

C. speciosa (Warder ex Barney) Warder ex Engelm. CN; flowers, D: 138, 165.

Catharanthus roseus (L.) G. Don synonym: *Vinca rosea* L. (Madagascar periwinkle) see text p. 141.

Caulophyllum thalictroides (L.) Michx. (blue cohosh) CN; see text p. 64.

Celastrus scandens L. (climbing bittersweet) CN; see p. 174.

Celtis spp. (hackberry) CN; pollen, AA: 109.

Cephalanthus occidentalis L. (buttonbush) CN; cephalanthin, cephalin (142); all parts toxic; bark, roots and leaves medicinal; bark, leaves toxic to livestock (127); I: 90, 126*, 127*, 142*.

Cercis canadensis (eastern redbud) CN; seedpods, seeds; I: 138s; 165.

Cestrum spp. (jessamine) C; see text p. 151.

C. diurnum L. (day jessamine) C; see text p. 151.

C. nocturnum L. (night-blooming jessamine) C; see text p. 151.

C. parqui L'Her. (willow-leaved jessamine) C; see text p. 151.

Chamaecyparis thyoides (L.) BSP (Atlantic white cedar) CN; extractable convulsant thujone; used as abortifacient; I: 106.

Chamedorea erumpens H. E. Moore (bamboo palm) C; juice of the fruit is very irritating (128).

Chamaelirium luteum (L.) Gray (devil's bit) CN; I: 143?.

Chamaesyce spp. (spurge) N; milky juice in all parts D: 142s.

C. hirta (L.) Millsp. N; see *Euphorbia hirta*.

C. hyssopifolia Small. synonym: *Euphorbia hyssopifolium* L. N; irritant milky sap; D: 128.

Chelidonium majus L. (celandine) CN; see text p. 69.

Chenopodium spp. CN; pollen, AA: 109; sap, P: 109.

C. album L. (lamb's quarters) CN; all parts with soluble oxalates, nitrates, HCN; I: 142s.

C. ambrosioides L. (Mexican tea) CN; see text p. 48.

Chimaphila umbellata (L.) W. Barton (pipsissiwa, rat's bane) CN; leaves, stem; D: 72, 165.

Chionanthus virginicus L. (fringe tree) CN; medicinal overdoses of the roots are toxic; I: 142s.

Chrysanthemum spp. C; lactones (109); irritant oleoresins (142); AA: 109; I: 138; D: 72, 106, 109, 142.

Cicer arietinum L. (chickpea) CN; seeds slightly cyanogenic; 0.8 mg/ 100 g tissue (145).

Cichorium intybus L. (chicory) CN; roots; D: 108.

Cicuta spp. CN; cicutoxin in all parts (106); I: 106d, 138d, 143d.

C. maculata L. (water hemlock) CN; see text p. 128.

C. mexicana Coulter & Rose N; I: 142d.

Ready Reference List

Cimicifuga racemosa (L.) Nutt. (rattleroot, bugbane) CN; roots medicinal, overdoses toxic; I: 113.

Cinnamomum camphora (L.) Presl. (camphor tree) C; safrole is a major component of the oil (67); camphor (142); AA: 109; I: 142; D: 109, 142.

C. zeylanicum Blume (cinnamon) C; safrole (100); phellandrene, a monocyclic terpene (109); D: 109.

Citrus spp. C; d-limonene or others in fruit peel oil; P: 109, 142, 165; D: 106, 109.

C. aurantifolia (L.) Swingle (lime) C; see text p. 92.

C. aurantium L. (sour orange) CN; see text p. 92.

C. limon (L.) Burm. f. (lemon) C; fruit peel oil; D: 106.

Clematis spp. (virgin's bower, leather flower) CN; see text p. 58.

Clematis virginiana L. CN; leaves; D: 72, 113.

Cleome spp. (spider flower) CN; seeds, sap, D: 165.

Clivia spp. C; lycorine and others; gastroenteric irritant; I: 106s, 138.

C. miniata Regel C; bulbs medicinal in Africa; lycorine and others; I: 142s.

C. nobilis Lindl. C; bulbs with alkaloids clivine, clivianine; I: 142, 171.

Clusia rosea Jacq. (balsam apple) CN; fruit, sap, gastroenteric irritant, chiefly to intestinal mucosa (106); bark, leaves, fruit medicinal (127); I: 72, 142.

Cnidoscolus spp. N; stinging hairs; roots edible and medicinal; I: 72, 109, 142, 165.

C. stimulosus (Michx.) Engelm. & Gray (spurge nettle) N; see text p. 97.

Cocculus laurifolius (Roxb.) DC. (laurel leaf, snailseed) C; leaves, seeds, bark, wood; I: 127?, 142s; leaves with cocculidine, coclifoline; seeds with unknowns; bark, wood with coclaurine, laurafoline, trilobine; all parts with coclamine, coclifoline and others (142).

Codiaeum spp. C; latex; D: 109.

C. variegatum (L.) Blume var. *pictum* (Lodd.) Mull. Arg. C; all parts; D: 142.

Coffea arabica L. (coffee) C; raw beans and smoke from them; D: 165; caffeine (109); novelty in south Florida when protected from the cold (127).

Colchium spp. C; all parts; colchicine and others; I: 142.

C. autumnale L. (autumn crocus) CN; see text p. 23.

Colocasia spp. (elephant's ear, dasheen) CN; gastroenteric irritant, mostly mouth and throat (106); calcium oxalate crystals (106); I: 49, 106, 138; D: 49, 106, 109, 138.

C. antiquorum (L.) Schott C; all parts; calcium oxalate crystals; I: 90, 138, 142, 165; D: 142.

C. esculentum (L.) Schott C; as above; I, D: 142.

Comptonia spp. (sweet fern) CN; pollen; AA: 109.

Conium maculatum L. (poison hemlock) CN; see text p. 130.

Consolida ambigua (L.) Ball & Heyw. synonym: *Delphinium ajacis* L. (rocket larkspur) CN; mostly young leaves and seeds; ajacine and others (142); I: 142sd; D: 72, 142, 165.

Convallaria majus L. (lily-of-the-valley) CN; see text p. 25.

Convulvulus spp. (bindweed) CN; poisoning as in morning glory (*Ipomoea*); I: 138.

Conyza canadensis (L.) Cronquist (horseweed) N; oleoresins (109, 142); pollen, AA: 109; leaves, D: 72, 106, 109, 142, 165.

Cooperia drummondii Herb. C; see *Zephyranthes brazosensis*.

Coriaria spp. C; see p. 175.

C. myrtifolia L. C; fruits with coriamyrtin (39); I: 106d.

Ready Reference List

Corylus spp. (hazelnut) CN; pollen, AA: 109.

Crataegus spp. (hawthorn) CN; fruit; I: 138, 165.

Crescentia cujete L. (calabash tree) C; fruit pulp purgative; large amounts have caused abortion in cattle (142).

Crinum spp. (milk & wine lily) CN; see text p. 35.

C. americanum L. (southern swamp crinum) CN; bulbs with lycorine and others; I: 90, 142s.

C. asiaticum L. CN; bulbs as above; I: 142s.

C. latifolium L. var. *zeylanicum* (L.) Hook. f. ex Trimen CN; bulbs as above; I: 142s.

Crotolaria spp. (rattlebox) CN; see text p. 79.

C. sagittalis L. N; all parts; I: 142*, 143.

C. spectabilis Roth. CN; all parts with monocrotoline and others; I: 109, 142*.

Croton spp. CN; croton oil in seeds, leaves, stem (142); I: 138d; D: 109, 138, 142s.

C. capitatus Michx. (woolly croton) N; see text p. 98.

C. tiglium L. (purging croton) CN; purgative oil (101); oil carcinogenic on mouse skin (21); contains phorbol (55); I: 101, 165sd; D: 165.

Cryptostegia grandiflora R. Br. (Palay rubbervine, purple allamanda) CN; see text p. 144.

C. madagascariensis Boj. (Madagascar rubbervine) CN; I: 127, 128, 142d; D: 72, 142.

Cucumis sativus L. (cucumber) CN; skin, mouth; D: 165.

Cucurbita lagenaria L. CN; see *Lagenaria*.

Cupressus spp. (cypress) CN; thujone, used as abortifacient; see *Juniperus* p. 7; I: 106.

Cuscuta spp. (dodder) N; all parts with unknowns; abortifacient; has caused nausea, vomiting, depression; I: 142s.

Cycas circinalis L. (false sago palm, crozier cycad) C; see text p. 3.

C. revoluta Thunb. C; cycasin (132); neocycasins A,B,C,E (133); male cones with noxious odor; AA: 142; I: 106, 128, 142d; D: 142.

Cyclamen spp. C; tubers with saponin cyclamin (71); I: 90, 138d, 165sd.

Cynara spp. (artichoke) C; lactones (108); D: 108, 109.

Cynodon dactylon (L.) Persoon (Bermuda grass) CN; all parts; cyanogenic glycosides that release HCN (109); pollen, AA: 109; D: 106, 109.

Cynoglossum spp. (hound's tongue) CN; irritant hairs, D: 109.

Cypripedium spp. (lady's slipper orchid) CN; see text p. 38.

C. acaule Ait. (pink lady's slipper) CN; D: 106.

C. calceolus L. (yellow lady's slipper) CN; D: 106.

C. candidum Muhl. ex Willd. (small white lady's slipper) CN; D: 106.

C. reginae Walter (showy lady's slipper) CN; D: 35, 106.

Cytissus laburnum L. CN; see text p. 81.

C. scoparius (L.) Link (Scotch broom) CN; seeds; quinolizidine alkaloid sparteine (109); I: 90.

Dactylis spp. (orchard grass) CN; pollen; AA: 109.

Daphne spp. C; all parts, usually berries; gastroenteric irritation, chiefly intestinal mucosa (106); I: 90d, 106d, 138d; D: 90, 106, 109.

D. mezereum L. C; see p. 175.

Datura spp. CN; I: 128, 138d, 142sd; D: 109, 142s.

D. candida (Pasq.) Saff. C; see *Brugmansia candida*.

D. innoxia Mill. C; all parts with atropine, scopolamine, hyoscyamine (106); used medicinally (128); I: 106.

D. metel L. (hairy thorn apple) CN; all parts as above (106, 142); used medicinally (128); I: 49d, 106, 127, 128, 135, 142d; D: 142.

D. stramonium L. (jimsonweed) N; see text p. 152.

D. suaveolens Humb. & Bonpl. ex Willd. C; see *Brugsmansia suaveolens*.

Daubentonia punicea (Cav.) Benth. CN; see text p. 86.

Daucus carota L. (Queen Anne's lace) N; see text p. 131.

D. carota L. var. *sativa* DC. (cultivated carrot) C; leaves with furocoumarins; P: 109, 142; D: 106, 165.

Delphinium spp. (larkspur) CN; see text p. 59.

D. ajacis L. CN; see *Consolida*.

Dicentra spp. (Dutchman's breeches) CN; see text p. 72.

D. canadensis (Goldie) Walpers N; I: 138.

D. cucullaria (L.) Bernh. CN; leaves, roots; the alkaloid bicuculline (39); I: 90, 138.

Dichondra repens J. Forster CN; D: 106.

Dictamnus albus L. (gas plant) CN; all parts with furocoumarins (55, 106); P: 106, 109, 138, 165; D: 72, 106.

Dieffenbachia spp. (dumbcane) C; see text p. 13.

D. maculata (Lodd.) G. Don synonym: *D. picta* Schott cv Rudolph Roehrs. C; I: 44s*.

D. seguine (Jacq.) Schott C; I: 72, 90, 101, 109, 128, 142sd; D: 90, 128, 142, 165.

Digitalis spp. C; leaves, flowers, seeds with digitoxin and others; I: 49.

D. purpurea L. (foxglove) C; see text p. 161.

Digitaria spp. CN; pollen, AA: 109.

D. sanguinalis (L.) Scop. (crabgrass) CN; D: 106, 109.

Dioscorea bulbifera L. (air potato) C; see p. 175.

D. villosa L. CN; I: 138.

Dirca spp. CN; sap; D: 109.

D. palustris L. (wicopy, leatherwood) CN; see text p. 122.

Distichlis spp. (saltgrass) N; pollen, AA: 109.

Dolichos lablab L. C; see *Lablab* on p. 178.

Dracaena sanderana Hort. Sander ex Mast C; leaf; I: 44s*.

Drosera spp. (sundew) CN; crushed leaves; D: 165.

D. rotundifolia L. (roundleaved sundew) CN; all parts medicinal, irritant; only in small amounts; I: 113, 138.

Duranta repens L. (golden drop) CN; see p. 175.

Echinocystis lobata (Michx.) Torr. & Gray (wild cucumber) CN; roots, seeds; I: 90.

Echites umbellata Jacq. (devil's potato) N; roots, milky sap with unknowns; I, D: 142s.

Echium spp. CN; pyrrolizidine alkaloids; I: 97.

E. vulgare L. (viper's bugloss) N; I: 129, 138d, 165sd.

E. plantagineum L. CN; pyrrolizidine alkaloid retrorsine; I: 109.

Epipremum aureum (Linden & Andre) Bunt. synonym: *Raphidiophora aurea* Birdsey C; see text p. 14.

Equisetum arvense L. (horsetail) N; see p. 176.

Eriobotrya japonica (Thunb.) Lindl. (loquat) CN; seeds with amygdalin that releases HCN (109, 142); unbroken seeds are harmless (106).

Eruca vesicaria (L.) Cav. synonym: *E. sativa* Mill. (rocket salad) CN; P: 165.

Ervatamia coronaria (Jacq.) Stapf. synonym: *Tabernaemontana divar-*

icata (L.) R. Br. (crape jessamine) C; see p. 176.

Erythrina spp. CN; seeds; I: 138, 165.

E. herbacea L. (eastern coral bean) CN; see p. 176.

E. variegata L. var. *orientalis* Merr. (coral tree) C; see p. 176.

Erythronium spp. (adder's tongue) CN; bulbs with the phytoalexin tulipalin A which causes "tulip fingers" (55); I: 138, 165; D: 55.

Eucalyptus spp. C; see text p. 123.

Eucharis grandiflora Planch. & Lind. (Amazon lily) C; roots with lycorine; I: 142s.

Euonymus spp. CN; bark, leaves, seeds; I: 72s.

E. americanus L. (hearts-a-bustin') CN; leaves, bark, seeds, roots; bark, roots medicinal; I: 142s.

E. atropurpureus Jacq. (strawberry bush) CN; see text p. 114.

E. europaeus L. (spindle tree) CN; see text p. 114.

Eupatorium spp. (boneset, dogfennel) CN; foliage with lactones; D: 106, 109.

E. perfoliatum L. (thoroughwort) CN; all parts with unknowns; medicinal overdoses toxic; I; 142s.

E. rugosum Houtt. N; see text p. 168.

Euphorbia spp. (spurge) CN; see text p. 99.

E. corollata L. (flowering spurge) CN; milky sap; I, D: 72.

E. cotinifolia L. (red spurge) C; all parts, especially the sap; I: 128, 142d; D: 127, 128, 142.

E. cyparissus L. (cypress spurge) CN; sap; I, D: 72.

E. esula L. (leafy spurge) CN; the ingenol ester ingenol-3-hexadecanoate that is carcinogenic on mouse skin (1).

E. hirta L. synonym: *Chamaesyce hirta* (L.) Millsp. N; sap with euphorbon and other irritants; D: 142.

E. hyssopifolium L. N; see *Chamaesyce hyssopifolium*.

E. ipecacuanhae L. synonym: *Tithymalopsis ipecacuanhae* (L.) Small. (wild ipecac) CN; roots medicinal, laxative, emetic; large doses toxic; I: 142.

E. lactea Haw. (candelabra cactus) CN; sap in all parts; injurious spines; I: 128, 142; D: 106, 127, 128, 142.

E. lathyris L. (caper spurge) CN; irritant sap; an ingenol ester ingenol-3-hexadecanoate that is carcinogenic on mouse skin (1); also the irritant ester 3-O-hexadecanoylingenol (55).

E. maculata L. (spotted spurge) N; I: 72; D: 72, 90, 106, 143.

E. marginata Pursh (snow-on-the-mountain) CN; I: 72, 90, 138, 143; D: 72, 90, 106

E. milii Ch. des Moul. (crown-of-thorns) C; very spiny; I, D: 72, 142.

E. pulcherrima Willd. (poinsettia) C; irritant latex, I: 49, 127; has caused deaths: 11, 90, 101, 135, 138, 142s; poisonous in quantities: 72, 109, 128; dermatitis: 49, 106, 127, 128, 135, 142, 171; nontoxic (44); see text p. 99.

E. tirucalli L. (pencil tree) C; irritant sap, D: 127, 135; I: 142d, 171d; oral inflammation (128); sap with irritant factors; 13-O-acetyl-12-O-acylphorbol and 12-O-acetyl-13-O-acylphorbol (59).

Fagopyrum spp. (buckwheat) CN; flour may cause bronchial asthma when ingested or inhaled by some people (91); AA: 109.

F. esculentum Moench. CN; see text p. 45.

F. sagittatum Gilib. synonym: *F. esculentum* Moench. CN; see text p. 45.

F. sagittatum (L.) Karsten CN; see text p. 45.

Fagus spp. (beech) CN; see p. 176.

F. grandiflora Ehrh. (American beech) CN; I: 72s, 142s.

F. sylvatica L. (European beech) C; pollen, AA: 90; nuts, I: 72s, 90.

Festuca spp. (fescue) CN; pollen, AA: 109.

Ready Reference List

Ficus spp. (fig) C; stems and leaves with furocoumarins, P: 142s, 165; psoralens in leaves, D: 91; proteolytic enzyme ficin; D: 55.

F. carica L. C; see text p. 39.

F. elastica Roxb. ex Horn. (rubberplant) C; I: 138; nontoxic (44).

F. pumila L. (creeping fig) C; latex in leaf and stem; P: 142; D: 106, 142.

Foeniculum vulgare Miller (fennel) CN; the phenol anethol; the oil also with ketones and terpenes; P, D: 109.

Fraxinus spp. (ash) CN; pollen oil with oleoresins; AA: 109; D: 106, 109, 142s.

Fritillaria meleagris L. (snakeshead fritillary) C; roots, heart depressant alkaloid, poisoning in Europe (109); I: 110, 134, 138, 165.

Gaillardia spp. (blanket flower) CN; all parts with lactones (109) and irritant oleoresins (142); D: 106, 109, 142, 165.

Gaultheria procumbens L. (wintergreen) CN; leaves medicinal; pure oil is irritant; D: 113.

Gelsemium rankii Small (yellow jessamine) N; all parts with gelsemine, sempervirine; I: 72, 142s; D: 142s.

G. sempervirens L. (Carolina jessamine) CN; see text p. 137.

Geranium spp. N; D: 106.

Geum rivale L. (water avens) CN; all parts medicinal; excessive amounts with unpleasant side effects (113).

Ginkgo biloba L. C; see text p. 6.

Gladiolus spp. C; corm with unknowns; I: 138, 142, 165.

Glechoma hederacea L. synonym: *Nepeta hederacea* Benth. (ground ivy) CN; herb medicinal and toxic in large amounts (113).

Gloriosa spp. (climbing lily) C; underground parts; I: 138, 165.

G. rothschildiana O'Brien C; all parts with colchicine (106, 142); I: 106d, 127, 128d, 138d, 142d.

G. superba L. (glory lily) C; see text p. 26.

Gossypium spp. (cotton) C; see p. 176.

Grevillea banksii R. Br. C; sap; I: 138; D: 72, 106, 138, 142s, 165.

G. robusta A. Cunningham (Australian silkoak) C; D: 106, 128, 142s.

Guaicum officinale L. (lignum vitae) C; see p. 177.

Gymnocladus dioica (L.) K. Koch (Kentucky coffee tree) CN; see text p. 80.

Haemanthus spp. (blood lily) C; bulbs with lycorine and others; some species medicinal and have caused poisoning and death; I: 142sd.

H. coccineus L. C; bulbs with lycorine, coccinine, manthidine and others; I: 142s.

H. multiflorus Martyn C; bulbs with haemanthidine and others; juice irritant to tongue and lips; I: 142.

Hedeoma pulegioides (L.) Pers. (pennyroyal); CN; leaves, flowering tops; oil with pulegone; overdoses are toxic; I: 109, 138; D: 109.

Hedera spp. CN; D: 109.

H. canariensis Willd. (Algerian ivy) C; all parts, especially leaves and berries; D: 106, 142s.

H. helix L. (English ivy) CN; see text p. 126.

Helenium spp. (sneezeweed); CN; see text p. 169.

H. amarum (Raf.) H. Rock synonym: *H. tenuifolium* Nutt. (bitterweed) CN; leaves, flowers, seeds with tenulin, a major sesquiterpene lactone (88); I: 101, 142, 165.

H. autumnale L. (bitterweed) CN; leaves, flowers and seeds with helenalin, a sesquiterpene lactone (55).

H. tenuifolium Nutt. CN; see *H. amarum*.

Heliotropium spp. (heliotrope) CN; see p. 177.

H. europaeum L. N; pyrrolizidine alkaloids heliotrine, lasiocarpine; I: 109.

Ready Reference List

Helleborus niger L. (Christmas rose); C; see text p. 60.

H. viridis L. CN; I: 138d.

Heracleum spp. CN; P, AA: 109; D: 106.

H. lanatum Michx. CN; fresh foliage; rootstock and seed medicinal (113); D: 90s, 113.

H. mantegazzianum Somm. & Levier (giant hogweed) CN; see p. 177.

H. sphondylium L. (hogweed) CN; furocoumarins; P: 164.

Hippeastrum spp. C; bulbs; I: 72, 101, 171.

H. equestre (Ait.) Herb. C; see *H. puniceum*.

H. puniceum (Lam.) Voss synonym: *H. equestre* (Ait.) Herb. C; bulbs; I: 109d.

H. vittata (L'Her) Herbert (Barbados lily) C; lycorine, tazzetine and others (142); I: 138, 142s; D: 142s.

Hippomane mancinella L. (manchineel) N; see text p. 101.

Holcus spp. (velvet grass) CN; pollen; AA: 109.

Humulus spp. (hops) CN; pollen; AA: 109.

H. lupulus L. CN; see text p. 43.

Hura crepitans L. (sandbox tree) CN; see text p. 102.

Hyacinth spp. (Hyacinth) C; calcium oxalate crystals (109); D: 106, 109.

H. orientalis L. (common hyacinth) C; see text p. 28.

Hydrangea spp. CN; see text p. 73.

H. arborescens L. (seven bark) CN; leaves and buds with hydrangin (142); I: 72, 142*; nontoxic (44).

H. macrophylla (Thunb.) Seringe (garden hydrangea) C; leaves and buds with hydrangin (142); HCN (106); I: 44s*, 49d, 72, 90d, 106, 138, 142.

H. quercifolia Bartram (oak-leaf or smooth hydrangea) CN; leaves, buds with hydrangin (142); I: 72, 142*?.

Hydrastis canadensis L. (golden seal) CN; see text p. 61.

Hymenocallis spp. (spider lily) CN; bulbs with lycorine and others; I: 128s, 142s*.

H. littoralis (Jacq.) Salisbury CN; bulbs with lycorine; I: 109, 128s, 142s*.

Hyoscyamus niger L. (henbane) CN; see text p. 154.

Hypericum spp. (St. John's wort) CN; some medicinal; leaves with hypericin (91); P: 91*, 109, 165?; I: 142s.

H. perforatum L. CN; all above ground parts medicinal; leaves with hypericin (91); P: 91*, 142*, 165; I: 138; D: 72.

Ilex spp. (holly) CN, see text p. 113.

I. aquifolium L. (English holly) CN; berries with illicin and others; gastroenteric irritant, chiefly to intestinal mucosa (106); I: 90sd, 106, 134, 165.

I. vomitoria Soland. in Ait. (yaupon) CN; see text p. 113.

Illicium anisatum L. (Japanese anise) CN; see p. 178.

I. floridanum Ellis CN; see p. 178.

Indigofera spp. (wild indigo) CN; seeds of some species toxic, some edible; all parts with the blue dye indigo, which is medicinal and toxic in large amounts (142); I: 138, 142s.

I. suffruticosa Miller N; medicinal; I: 135*, 142s.

Ipomoea spp. (morning glory) CN; seeds and roots of some are medicinal; large amounts may be toxic (142); I: 138, 142s, 165.

I. alba L. synonym: *Calonyction aculeatum* (L.) House (Moonflower, moonvine) CN; see p. 178.

I. pes-caprae (L.) R. Br. (beach morning glory) N; all parts medicinal, toxic in large amounts with nausea, diarrhea (142) I: 127, 142s.

I. purpurea (L.) Roth (tall morning glory) CN; seeds in large amounts; I: 90.

Ready Reference List

I. tricolor Cav. synonym: *I. violacea* L. C; see p. 145.

I. violacea CN; see text p. 145.

Iris spp. CN; see text p. 37.

I. germanica L. CN; orris oil, irone; D: 109.

I. pallida Lam. CN; as above; D: 109.

I. versicolor L. CN; see text p. 37.

Iva spp. (marsh elder) N; AA, D: 109.

I. angustifolia Nutt. ex DC N; pollen; D: 109.

I. xanthifolia Nutt. N; pollen, D: 106; allergic contact dermatitis (165).

Jacaranda spp. (green ebony) CN; D: 165.

Jacquinia spp. (barbasco, cudjoewood) C; all parts; I: 142s?.

J. barbasco (Loefl.) Mez C; all parts; I: 127?.

Jatropha spp. C; all parts with curcin and others (106, 142); I: 72, 138d, 142s; D: 106, 165.

J. curcas L. (physic nut) CN; see text p. 103.

J. gossypifolia L. (bellyache bush) CN; I: 72, 106, 127, 128, 135, 142d; D: 142.

J. integerrima Jacq. (peregrina) CN; I: 72, 127, 142; D: 142s.

J. multifida L. (coral plant) CN; I: 72, 106, 127, 128, 135, 138, 142d; D: 142.

J. podagrica Hooker CN; I: 127; D: 142s.

Juglans nigra L. (black walnut) C; stem and fruit hull with juglone, a naphthoquinone (166) which is a plant growth inhibitor (126); AA: 109; D: 90, 106, 142.

Juncus spp. (rush) CN; pollen; AA: 109.

Juniperus spp. (juniper, red cedar) CN; see text p. 7.

J. silicola (Small) Bailey (southern red cedar) CN; D: 142s.

J. virginiana L. (eastern red cedar) CN; leaves; I: 72, 106, 142s*; D: 72, 138, 142s, 165.

Kalanchoe pinnata (Lam.) Pers. (air plant) CN; unknowns; juice medicinal (127); I: 142s?.

Kalmia spp. (laurel) CN; all parts; I: 72d, 138s, 143d.

K. angustifolia L. (lambskill) CN; all parts; andromedotoxin (106); I: 72d, 106.

K. hirsuta Walter (wicky) CN; all parts; I: 72d, 142s?.

K. latifolia L. (mountain laurel) CN; see text p. 133.

K. poliifolia Wangenh. CN; I: 72d.

Lablab niger Medik. synonym: *Dolichos lablab* L. (hyacinth bean) CN; see p. 178.

Laburnum anagyroides Medik. (golden chain) CN; see text p. 81.

Lachnanthes caroliniana (Lam.) Dandy (bloodwort) CN; leaves, flowers, stems, roots medicinal, toxic in overdoses; I: 142.

Lactuca sativa L. (milkweed) CN; all parts; D: 106, 165.

Lagenaria siccraria S. Wats. synonym: *Cucurbita lagenaria* L. CN; seeds very toxic; fruit pulp has been used for roundworm; two children killed in Cuba; I: 109d.

Lagerstroemia indica L. (crepe myrtle) CN; bark, leaves, seeds, flowers medicinal, may be toxic in large amounts; I: 142s.

L. speciosa (L.) Pers. (crepe flower) C; as above; I: 142s.

Lantana camara L. CN; see text p. 146.

Laportea spp. N; stinging hairs; D: 109.

L. canadensis (L.) Weddell (wood nettle) N; see *Urtica* p. 41.

Larix decidua Mill. synonym: *L. europaea* DC. (European larch) C; bark, resin, needles medicinal, overdoses toxic; I: 113.

L. europaea DC. C; see *Larix decidua*.

Lathyrus spp. (sweet pea) CN; see text p. 82.

L. hirsutus L. (wild winter pea) CN; seeds with β-(gamma-L-glutamyl)-aminopropionitrile (142*).

L. odoratus L. C; see text p. 82.

L. pusillus Elliott (everlasting pea) C; I: 142*.

L. sativus L. C; see text p. 82.

Lecythidaceae (family of tropical trees); nuts of some species are toxic, some edible (72).

Ledum spp. (Labrador tea) CN; see *Kalmia* p. 133; I: 113.

L. groenlandicum Oeder (marsh tea) CN; herb medicinal, overdoses toxic; I: 113.

Leonotis nepetifolia (L.) R. Br. in Ait. (hollowstalk) CN; leaves medicinal (142); pollen, AA: 109, 135, 142; D: 109.

Leonurus cardiaca L. (motherwort) CN; leaves, oil; D: 72, 129, 165.

Lepidium sativum L. (garden cress) CN; seeds; D: 165.

Leucothoe spp. (fetterbush) CN; andromedotoxin (106, 142); see *Kalmia* p. 133; I: 106, 142s.

L. axillaris (Lam.) D. Don CN; as above; I: 142s.

L. populifolia (Lam.) Dipel N; see *Agarista*.

L. racemosa (L.) Gray CN; as above; I: 142s.

Ligustrum spp. (privet) CN; fruit with ligustrin, syringin and others (142); flowers may cause respiratory irritation when in bloom (109, 128, 142); I: 49d, 90d, 128d, 138d; D: 142s.

L. vulgare L. CN; see text p. 136.

Lilium tigrinum Ker-Gawl (tiger lily) CN; I: 138, 165.

Linaria vulgaris Mill. (toadflax) CN; all parts medicinal, even 20 drops may cause poisoning (113); I: 113, 138.

Linum usitatissimum L. (flax) CN; see text p. 91.

Liquidambar styraciflua L. (sweetgum) CN; storesins and others; pollen, AA: 109s; D: 109.

Lobelia spp. (cardinal flower, Indian tobacco) CN; see text p. 166.

L. cardinalis L. CN; lobeline (106); I: 49sd, 106.

L. inflata L. CN; lobeline (109); medicinal (72); I: 72d, 90, 106, 138, 142d; D: 72, 109, 142.

L. siphilitica L. (great lobelia); CN; I: 106, 138.

Lolium spp. (rye grass) CN; pollen, AA: 109.

L. temulentum L. (darnel); CN; see p. 178.

Lonicera spp. (honeysuckle) CN; berries; gastroenteric irritation (106); I: 106d, 138, 142s, 165.

L. japonica Thunb. (Japanese honeysuckle) CN; see text p. 163.

L. sempervirens L. (trumpet honeysuckle) CN; I: 142s.

L. tatarica L. CN; I: 138s.

Lotus spp. CN; cyanogenic glycosides; I: 109, 138d, 165sd.

Lupinus spp. (lupine) CN; see text p. 82.

L. perennis (wild lupine) CN; angustifoline, hydroxylupanine, lupanine (142); I: 143.

Lycium halimifolium Miller (matrimony vine) CN; all parts with atropine, hyoscyamine, scopolamine (106); I: 106, 138s, 142s*, 165.

Lycopersicon esculentum Miller (tomato) C; see text p. 155.

Lycopodium clavatum L. (running clubmoss) CN; all parts are toxic, except the spores which are used medicinally (113).

Lyonia spp. CN; all parts with andromedotoxin; see *Kalmia* in text page 133; I: 106.

L. ligustrina (L.) DC. (staggerbush) CN; as above; I: 142s.

L. lucida (Lam.) Koch CN; said to be nontoxic (142).

L. mariana (L.) D. Don CN; as in *Lyonia spp.*; I: 142s.

Macadamia ternifolia Muell. (macadamia nut) C; young leaves with high concentrations of HCN (142); I: 138, 142s.

Maclura spp. CN; milky sap; D: 109.

M. pomifera (Raf.) Schneider (Osage orange) CN; thick pulp may cause choking (90); pollen, AA: 109 D: 72, 90, 109, 129, 142, 165.

Magnolia spp. CN; D: 109.

M. grandiflora L. (southern magnolia) CN; D: 106.

Malus spp. (apple) CN; seeds with amygdalin; cyanogenic in large amounts; I: 72, 90d, 101, 138d, 165.

M. angustifolia (Ait.) Michx. (southern wild crabapple) CN; seeds as above; I: 142s.

M. sylvestris Mill. (apple) C; see text p. 74.

Mangifera indica L. (mango) C; see text p. 108.

Manihot spp. C; seeds with HCN (109); harmless unless broken (106); I: 106, 109.

M. esculenta Crantz (cassava) CN; see p. 178.

M. ultissima Pohl C; see *Manihot* on p. 179.

Manilkara spp. CN; milky sap; D: 109.

M. zapota (L.) VanRoyen (sapodilla) CN; see p. 179.

Melaleuca quinquenervia (Cav.) S. T. Blake (cajeput) C; see text p. 124.

Melanthium virginicum L. synonym: *Veratrum virginicum* (L.) Ait. CN; leaves, seeds; I: 142s.

M. woodii (Rob. ex Wood) Bod. synonym: *Veratrum woodii* Robbins (varebell) N; all parts; I: 142s.

M. parviflorum (Michx.) S. Wats. synonym: *Veratrum parviflorum* (Michx.) S. Wats. N; I: 72.

Melia azedarach L. (chinaberry) CN; see text p. 95.

Melilotus alba Desr. (white melilot) CN; all parts medicinal, large doses are toxic; I: 113.

M. officinalis (L.) Pall. CN; as above; I: 113.

Melothria pendula L. (creeping cucumber); N; I: 143?.

Menispermum canadense L. (moonseed) N; see text p. 67.

Mentha cardiaca (Gray) Baker CN; see *M. x gentilis*.

M. x gentilis L. synonym: *M. cardiaca* (Gray) Baker (scotch spearmint) CN; the ketone carvone; D: 109.

M. pulegium L. (European pennyroyal) CN; see p. 179.

M. spicata L. (spearmint) CN; spearmint oil with phellandrene and carvone; D: 109.

Metopium toxiferum (L.) Krug & Urban (poisonwood) CN; see text p. 109.

Mimosa spp. CN; seeds and other parts; some species with mimosine, a toxic amino acid (97); I: 128, 165.

M. pudica L. (sensitive plant) CN; seeds with mimosine, roots with unknowns (142, 167); the latter have been used for toothache in Panama, but cause vomiting if swallowed (142).

Mirabilis spp. (four o'clock) CN; I: 143.

M. jalapa L. CN; see text p. 49.

Momordica spp. CN; fruit; I: 138, 165.

M. balsamina L. (balsam apple) CN; see text p. 165.

M. charantia L. (balsam pear) CN; see text p. 165.

M. charantia L. var. *abbreviata* Ser. (wild balsam pear) CN; see text p. 165.

Monstera spp. C; stem, leaves with calcium oxalate crystals; gastroenteric irritant, chiefly mouth and throat; AA: 106; I, D: 49, 106.

M. deliciosa Liebm. (split-leaf philodendron) C; see text p. 15.

Moringa oleifera Lam. (ben tree) C; fresh, crushed leaves, large amounts of the pungent roots; I: 142s; D: 109, 142.

Morus spp. (mulberry) CN; pollen, AA: 109.

M. alba L. (white mulberry) CN; see text p. 40.

M. rubra L. (red mulberry) CN; see text p. 40.

Mucuna spp. CN; hairs with proteolytic enzyme; D: 109.

M. deeringiana (Bort) Merrill synonym: *Stizolobium deeringianum* Bort (velvet bean) CN; see p. 179.

M. pruriens (L.) DC. (cowhage) N; minute stinging hairs on pods with mucunain, serotinin and unknowns (142).

M. sloanei Fawc. & Rendle (donkey-eye) N; as above; D: 142.

Myrica spp. (wax myrtle) CN; pollen, AA: 109.

Myristica fragrans Houtt. (nutmeg) C; see p. 179.

Narcissus spp. C; see text p. 36.

N. jonquilla L. (jonquil) C; see text p. 36.

N. poeticus L. (poet's narcissus) C; see text p. 36.

N. pseudonarcissus L. (daffodil) C; see text p. 36.

Nasturtium officinalis R. Br. (watercress) CN; leaves, young shoots, roots medicinal; excessive use may be harmful; I: 113.

Nepeta cataria L. (catnip) CN; seeds; I: 138, 165.

N. hederacea Benth. CN; see *Glechoma*.

Nerium spp. L. (oleander) CN; D: 109.

N. oleander L. CN; see text p. 142.

Nicandra spp. (apple-of-Peru) CN; I: 134, 138, 165.

N. physalodes (L.) Gaertner (shoofly plant) CN; all parts except ripe fruit; roots with hygrine and others (142); I: 72, 142s, 143.

Nicotiana spp. (tobacco) CN; all parts with nicotine and other alkaloids (97, 109); I: 90d; D: 142s.

N. glauca R. Graham (tree tobacco) C; all parts with nicotine and other alkaloids (49); I: 49sd, 72d, 138, 142d; D: 142.

N. tabacum L. CN; see text p. 156

Ochrosia elliptica Labill (ochrosia plum) C; red fruits are toxic; I: 72, 127, 142s?, 165.

Ornithogalum umbellatum L. (star-of-Bethlehem) CN; see text p. 29.

Orontium aquaticum L. (golden club) CN; see text p. 16.

Oryza sativa L. (rice) C; D: 106.

Ostrya spp. (ironwood) CN; pollen, AA: 109.

Oxalis spp. (wood sorrel) CN; see p. 180.

Pachyrrhizus erosus (L.) Urban (yam bean) CN; leaves and seeds with pachyrrhizid; seeds also with rotenone and others (142); gastroenteric irritation, chiefly to intestinal mucosa (106); I: 106, 109, 142.

Paeonia spp. (peony) C; juice, I: 101.

P. officinalis L. C; rootstock medicinal; all parts toxic; flower tea can be fatal (113).

Papaver spp. (poppy) CN; all parts with isoquinoline alkaloids except seeds of some species (142); I: 90s, 142s?.

P. somniferum L. (opium poppy) CN; see text p. 70.

Parthenium hysterophorus L. N; sesquiterpene lactones; severe problems seen in Indian field workers with prolonged exposures; D: 14.

Parthenocissus quinquefolia (L.) Planchon (Virginia creeper) CN; see text p. 120.

Pastinaca sativa L. (wild parsnip) CN; see text p. 132.

Pedilanthus spp. CN; D: 109.

P. tithymaloides (L.) Poit. (redbird-cactus) CN; see p. 179.

Ready Reference List

Peltandra virginica (L.) Kunth. (arrow arum) CN; I: 143.

Pernettya spp. C; berries cause drunkenness, sensation of cold and paralysis (109); andromedotoxin (106); see *Kalmia* in text p. 133.

Petiveria alliacea L. (Guinea hen weed) N; all parts medicinal; causes bad flavors in milk if grazed by dairy animals; I: 142s*.

Peucedanum spp. (masterwort) N; P: 109.

P. ostruthium (L.) Koch (hogfennel) N; furocoumarins; P: 165.

Phacelia spp. (scorpion weed) CN; leaves; D: 72, 109, 165.

P. whitlavia Gray (California bluebells) CN; D: 106.

Phaseolus spp. CN; beans cyanogenic (109); I: 109; D: 165.

P. coccineus L. (scarlet runner bean) CN; beans; I: 72, 165.

P. lunatus L. (Java bean) C; see p. 180.

Philadelphus spp. (mock orange) CN; fruit; I: 138d, 165sd.

Philodendron spp. C; see text p. 17.

Phleum spp. (Timothy) C; pollen, AA: 109.

Phoradendron spp. N; I: 90d; D: 109.

P. serotinum (Raf.) M. C. Johnston (mistletoe) N; see text p. 44.

Physalis spp. (ground cherry) CN; leaves, unripe fruit with solanine (106); delayed gastroenteric irritation (106); I: 72, 106, 142s?, 143.

Phytolacca americana L. (pokeberry, poke salet) CN; see text p. 50.

P. rigida Small CN; I: 72, 142d.

Pieris spp. (fetterbush) CN; andromedotoxin (106); see *Kalmia* in text p. 133; I: 106.

P. phillyreifolia (Hooker) DC. (climbing heath) CN; see *Kalmia* in text p. 133; I: 142s.

Pilea cadierei Gagnep. and Guillaum. C; leaf, stem; I: 44s*.

Pinus spp. (pine) CN; pinene, a dicyclic terpene, sesquiterpenes cadinene, selinene and zingiberine; D: 109.

Piscidia piscipula (L.) Sargent (Jamaica dogwood) CN; leaves, twigs, bark, roots with rotenone; medicinal, toxic in large amounts (142); used in arrow poisons (142), to stupefy fish (127); I: 127*, 138, 142s.

Pistia stratiotes L. (water lettuce) CN; see text p. 18.

Pisum sativum L. (common pea) C; seeds are slightly cyanogenic (2.3 mg/100 g) (145).

Pithecellobium dulce (Roxb.) Benth. (monkey pod) CN; sap, bark, D: 72, 165; seeds, leaves, bark, roots medicinal (142); has sharp spines; aril is edible (127, 142).

P. saman (Jacq.) enth. C; see *Samanea*.

Pittosporum spp. C; leaves, stems, fruit; I: 90.

Plantago spp. (plantain) CN; pollen, AA: 109.

Platanus spp. (sycamore) CN; pollen, AA: 109.

Plumbago auriculata Lam. CN; see *P. capensis*.

P. capensis Thunb. synonym: *P. auriculata* Lam. CN; I: 127, 142, 165sd; D: 44s*, 72, 142s, 165.

P. indica L. C; see p. 180.

P. scandens L. CN; I: 142s.

Plumeria spp. (temple flower) C; latex, bark, rootbark; I: 142s; D: 72, 109, 138, 142s, 165.

P. rubra L. (nosegay) C; latex with calcium salts of plumeric acid, serotinic acid and others; bark with the glycoside plumierid; I, D: 142s.

Poa spp. (bluegrass) CN; pollen, AA: 109.

Podocarpus macrophylla (Thunb.) D. Don (Japanese yew) C; see text p. 9.

Podophyllum peltatum L. (mayapple) CN; see text p. 65.

Poinciana gilliesii Hook. CN; see *Caesalpinia* on p. 174.

Polygala senega L. (Seneca snakeroot) CN; underground parts; I: 101, 165.

Polygonatum spp. (Solomon's seal) CN; fruit with anthraquinone (165); causes vomiting, diarrhea (109); I: 109, 134, 165.

Polygonum spp. (smartweed) CN; acrid juice in all parts; some medicinal and eaten as a potherb (142); D: 72, 109, 142s, 165.

P. hydropiper L. (marsh pepper) N; acrid juice; medicinal; I, D: 113.

P. punctatum Elliott (dotted smartweed) N; leaves with about 7% calcium oxalate (109); medicinal (13); I: 109, 113; D: 113.

Polyscias spp. (aralia) C; see text p. 127.

P. balfouriana (Hort. Sander) Bailey (geranium-leaf aralia) C; foliage with saponins (142); I: 142s; D: 128, 142s.

P. guilfoylei (Bull.) Bailey (gallego) C; foliage with saponins (142); I: 142s; D: 127, 128, 142s.

Poncirus trifoliata (L.) Raf. (trifoliate orange) CN; see text p. 93.

Pongamia pinnata (L.) Pierre (pongam) C; see p. 180.

Populus spp. (cottonwood) CN; oleoresins (109); pollen, AA: 109; D: 106, 109, 142s.

Portulaca oleracea L. (purslane) CN; see p. 180.

Pothos aureus Lind. & Andre C; see text p. 14.

Primula spp. (primrose) CN; hairs on leaves and stem with the quinone primin (109, 142); D: 106, 109, 142s.

P. obconica Hance C; see text p. 135.

Prunus spp. CN; see text p. 75.

P. americana Marshall (plum) CN; broken or crushed seeds; I: 49.

P. amygdalus Batsch. var. *amara* (DC) Focke (bitter almond) C; see text p. 75.

P. armeniaca L. (apricot) C; amygdalin in seeds (145); I: 49, 142d, 145.

P. caroliniana (Mill.) Ait. (cherry laurel) CN; all parts except fruit pulp; I: 49.

P. pennsylvanica L.f. (pin cherry) CN; all parts except fruit pulp; I: 90.

P. persica (L.) Batsch. (peach) C; seeds with amygdalin (145); I: 126, 145.

P. serotina Ehr. (wild or black cherry) CN; all parts except fruit pulp; bark and dried fruit medicinal (72); I: 49, 72d, 90, 126, 138, 142.

P. virginiana L. (choke cherry) CN; all parts except fruit pulp (49); I: 49, 90.

P. x cerasus L. (cultivated cherry) C; all parts except fruit pulp; I: 49.

Pseudocalymma alliaceum Sandw. (garlic vine) C; all parts with volatile odors, AA: 128, 142s.

Psoralea spp. (scurf pea) C; seeds; P: 55; D: 106; seed extracts from *P. corylifera* were used in ancient India to re-pigment white patches of skin (61).

P. argophylla Pursh N; seeds; I: 101, 165.

Pteridium aquilinum (L.) Kuhn (bracken) CN; see p. 180.

Pulsatila patens (L.) Mill. synonym: *Anemone patens* L. (prairie crocus) N; irritant sap; D: 106, 165.

Punica granatum L. (pomegranate) C; stem and rootbark with the tropane alkaloid isopelletierine (109), tannic acid and unknowns (142); medicinal overdoses may be toxic (113, 142); I: 113, 138, 142s.

Pyracantha spp. (firethorn) CN; berries; I: 138, 165.

P. coccinea Roem. C; as above; I: 109, 142; D: 142.

Pyrus spp. (pear) C; see *Malus* on text p. 74.

Quercus spp. (oak) CN; leaves, seeds with tannin; this should be leached out before being eaten (72s, 138, 142s); large amounts over extended periods are toxic (113, 165).

Ready Reference List

Quisqualis indica L. (Rangoon creeper) C; seeds medicinal; more than 4–5 are toxic (142); leaves, fruit also medicinal (127); I: 127, 142s.

Ranunculus spp. (crowfoot, buttercup) CN; see text p. 62.

R. acris L. (tall field buttercup) CN; I: 90; D: 90, 106.

R. bulbosus L. (bulbous buttercup) C; D: 106.

R. scleratus L. (cursed crowfoot) N; D: 106.

Raphanus raphanistrum L. (wild radish) N; see p. 181.

R. sativus L. (radish) C; see p. 180.

Raphidiophora aurea Birdsey C; see text p. 14.

Rhamnus spp. (buckthorn) CN; see text p. 118.

Rheum rhaponticum L. (rhubarb) c; see text p. 46.

Rhododendron spp. CN; see text p. 134.

Rhodotypos spp. (jetberry bush) C; fruit with cyanogenic glycosides that release HCN (109); I: 109, 138d, 165.

R. scandens (Thunb.) Makino synonym: *R. tetrapetala* Makino C; see p. 181.

R. tetrapetala Makino CN; see *Rhodotypos* in notes p. 161.

Rhoeo spp. CN; D: 109.

R. spathacea (Swartz) Stearn (oyster plant) CN; see text p. 20.

Ricinus communis L. (castor bean) CN; see text p. 104.

Rivea corymbosa (L.) Hallier CN; see *Turbina*.

Rivina humilis L. (bloodberry) CN: all parts; gastroenteric irritation, chiefly intestinal mucosa (106); toxicity similar to *Phytolacca* (106, 128, 142); leaves medicinal (142); fruit is toxic (128); I: 106, 128, 138, 142.

Robinia pseudoacacia L. (black locust) CN; see text p. 85.

R. viscosa Vent. (clammy locust) CN; see text p. 85.

Rosa spp. (rose) C; pollen, AA: 109; allergic contact dermatitis after contact with the flowers (109).

R. centifolia L. (cabbage rose) C; rose oil; D: 109.

R. gallica L. (French rose) C; rose oil; D: 109.

R. odoratus L. (tea rose) C; D: 106.

R. x damascena Mill. (damask rose) C; rose oil; D: 109.

Rosmarinus officinalis L. (rosemary) C; all parts medicinal, overdoses may be fatal (113); I: 113d, 138; D: 109.

Rudbeckia hirta L. (black-eyed Susan) CN; leaves with irritant oleoresins; D: 106, 142.

Rumex spp. (dock, sorrel) CN; see text p. 47.

R. acetosa L. (sand begonia, garden sorrel) CN; delayed gastroenteric irritation caused by oxalic acid (106); eaten when properly cooked; I: 90, 106, 138; D: 90.

Ruta graveolens L. (rue) CN; see text p. 94.

Salicornia bigelovii Torr. (glasswort) N; leaves; edible if cooked; I: 142s.

S. virginica L. N; as above; I: 142s.

Salix spp. (willow) CN; pollen, AA: 109.

Salvia officinalis L. (sage) C; leaves medicinal; excessive or prolonged use may be harmful (113).

Samanea saman (Jacq.) Merrill synonym: *Pithecellobium saman* (Jacq.) Benth. (monkey pod, raintree) C; saponins; gastroenteric irritation, chiefly intestinal mucosa (106).

Sambucus spp. (elderberry) CN; see text p. 164.

S. canadensis L. CN; roots, bark, stem, leaves with cyanogenic glycosides and alkaloids (49, 142); slightly in flowers, berries; flowers medicinal (142); I: 49, 72, 90, 126, 142.

S. canadensis L. var. *luciniata* Gray synonym: *S. simpsonii* Rend. (gulf elder) N; as above; I: 72, 142.

S. racemosa L. (red-berried elder) CN; I: 72.

S. simpsonii Rend. N; see *S. canadensis*.

Sanguinaria canadensis L. (bloodroot) CN; see text p. 71.

Sapindus spp. (soapberry) CN; fruits; D: 109, 165.

S. saponaria L. var. *drummondii* (Hook. & Arn.) L. Benson CN; fruit, bark, roots, seeds (106, 142s); with saponins (106); gastroenteric irritation, chiefly intestinal mucosa (106); leaves, seeds, roots medicinal (142); fruit used to poison fish (127); I, D: 106, 142s.

Sapium sebiferum (L.) Roxb. (popcorn tree) CN; see text p. 106.

Saponaria officinalis L. (bouncing bet) N; see text p. 53.

Sassafras albidum (Nutt.) Nees. CN; sassafras oil (109); one major component is safrole, a weak carcinogen in rats (75); D: 109s, 126.

Schinus terebinthifolius Raddi (Brazilian pepper) CN; see text p. 110.

Scindapsus aureus Engl. C; see text p. 14.

Secale cereale L. (rye) C; pollen, AA: 109.

Sedum acre L. (mossy stonecrop) CN; sap; D: 129, 165.

Senecio spp. (squaw weed) CN; see text p. 170.

S. aureus L. CN; used by Indians to induce abortion and childbirth (109).

S. confusus (DC.) Britten (Mexican flame vine) CN; see text p. 170.

S. glabellus Poiret (butterweed) N; all parts; I, D: 142s.

S. jacobea L. (tansy ragwort) CN; see text p. 170.

Sesbania drummondii (Rydberg) Cory (rattlebush) CN; seeds; I: 49d, 142.

S. grandiflora Persoon (corkwood tree) CN; seeds?; bark medicinal; I: 142s?.

S. punicea (cav.) Benth. synonym: *Daubentonia punicea* (Cav.) Benth. (purple sesbane) CN; see text p. 86.

S. vesicari (Jacq.) Elliott (bagpod) N; seeds; I: 49, 142.

Setcreasea purpurea Boom CN; see text p. 21.

Sisymbrium irio L. (London rocket) N; sap from seeds; D: 165.

Solandra spp. (trumpet flower) CN; see text p. 158.

S. grandiflora Swartz. C; all parts except ripe fruit; I: 128, 142s; D: 142s.

S. guttata D. Don ex Lindl. C; as above; I: 128, 142s; D: 142s.

S. longiflora Tussac C; as above; I: 128, 142s; D: 142s.

S. maxima (Sesse & Moc.) Greene synonym: *S. nitida* Zucc. C; as above; I: 127, 128, 142s; D: 142s.

S. nitida Zucc. C; see *S. maxima*.

Solanum spp. (nightshade) CN; see text p. 159.

S. aculeatissimum Jacq. (loveapple) CN; all parts with solanine, causing delayed gastroenteric irritation (106, 142).

S. americanum Miller (common nightshade) N; all parts except possibly ripe berries with solasonine and others (142); I: 49, 72d, 142d.

S. carolinense L. (horse nettle) CN; all parts with solanine and others (142); I: 49, 72d, 106, 142sd, 143.

S. dulcamara L. (bittersweet) CN; berries with solanine; delayed gastroenteric irritation (106); I: 72d, 106sd, 138.

S. eleagnifolium Cav. (silver nightshade) N; prickly; solasonine and others in all parts (142); I: 49, 142*.

203

S. gracile Link (graceful nightshade) N; solanine causing delayed gastroenteric irritation (106); I: 106, 142sd.

S. intrusum Sorca C; see *S. melanocerasum*.

S. melanocerasum All. synonym: *S. intrusum* Sorca (garden huckleberry) C; all parts; unripe berries very toxic with solanigrine and others; I: 142sd.

S. melongena L. (eggplant) C; all parts with solanine and others; less in raw fruit; all parts medicinal and widely cultivated for the edible fruit; I: 142s.

S. nigrum L. (black nightshade) CN; all parts except possibly ripe berries with solanine and others (106, 142); ripe berries are made into jam (142); I: 90d, 128d, 142d.

S. pseudocapsicum L. (Jerusalem cherry) C; all parts, especially berries with solanine and others, causing delayed gastroenteric irritation (106); I: 44s*, 49d, 72d, 90, 101, 128d, 138, 142d, 171.

S. seaforthianum Andrews (Brazilian nightshade) C; all parts (142); unripe fruit (127); I: 127s, 128, 142.

S. tuberosum L. (potato) C; see text p. 159.

Solidago spp. (goldenrod) CN; pollen, AA: 109; D: 109.

Sorbus spp. (mountain ash) CN; fruit; I: 134, 165.

Sorghum spp. CN; pollen, AA: 109.

Spigelia anthelmia L. (West Indian pinkroot) N; all parts with spigiline and others; medicinal overdoses are toxic; I: 142d.

S. gentianoides Chapman N; as above; I: 142s.

S. marilandica L. (Indian pink) CN; see text p. 139.

Spinacia oleracea L. (spinach) C; leaves with soluble oxalates; a very large amount may be toxic (142); I: 142s; D: 165.

Spiraea spp. (bridal wreath) CN; allergic reaction after contacting the flowers (109s).

Sterculia spp. C; irritant hairs in the fruits; D: 109.

S. apetala (Jacq.) Karsten (Panama tree) C; pods with irritant hairs; seeds medicinal; D: 142.

S. foetida (bangor nut) C; raw seeds, diarrhea; edible if cooked; I: 127, 142s.

Stillingia sylvatica L. (yaw root) CN; all parts (113); fresh root (142); with stillingine and others (142); cyanogenic glycosides that release HCN (109); small amounts medicinal, overdoses toxic (126); I: 113, 126, 142.

Stizolobium deeringianum Bort N; see *Mucuna* on p. 179.

Suaeda linearis (Elliott) Moq. (sea blite) N; leaves; edible if cooked in 2–3 waters; I: 142s.

Symphoricarpos albus (L.) Blake (snowberry) C; berries; gastroenteric irritation, chiefly to intestinal mucosa (106).

Symphytum officinale L. (common comfrey) CN; contains pyrrolizidine alkaloids found to be carcinogenic in rats and other laboratory animals (78).

Symplocarpus foetidus (L.) Salisb. (skunk cabbage) CN; see text p. 19.

Synadenium grantii Hook. f. (African milkbush) C; sap is very irritant to skin and eyes (128).

Syringa spp. (lilac) C; pollen, AA: 109.

Tabernaemontana divaricata (L.) R. Br. C; see *Ervatamia* on p. 176.

Tagetes spp. (marigold) C; all parts with lactones (109); D: 109, 165.

T. minuta L. (wild marigold) C; D: 108.

Tanacetum vulgare L. (tansy) CN; see text p. 172.

Taraxacum spp. (dandelion) CN; some people experience allergic reactions after coming in contact with the flowers (109).

Taxus spp. (yew) CN; see text p. 8.

T. baccata L. (English yew) C; taxine A & B; gastroenteric irritant, chiefly to intestinal mucosa (106); I: 72, 106d, 138.

T. canadensis Marshall (American yew) CN; as above; I: 106d, 143d.

T. cuspidata Sieb. & Zucc. (Japanese yew) C; as above; I: 72, 142s.

T. floridana Chapm. (Florida yew) CN; as above; I: 72, 142s.

Tephrosia spp. (goat's rue) CN; seeds with tephrosin (165); I: 165sd.

T. virginiana (L.) Persoon CN; see text p. 87.

Terminalia muelleri Benth. (Mueller's terminalia) C; not very common, but volatile emanations may cause respiratory difficulties, especially in bloom (128).

Tetragonia tetragonioides (Pallas) Kuntze synonym: *T. expansa* Murr. (New Zealand spinach) C; leaves with soluble oxalates; toxic if eaten raw and in large amounts (142s).

Tetrapanax papyriferus Koch (rice paper plant) C; pollen with unknowns, D: 142.

Thalictrum spp. (meadow rue) CN; all parts; D: 165.

Thevetia peruviana (Pers.) K. Schumann. (yellow oleander) C; see p. 181.

Thuja occidentalis L. (American arborvitae) C; see p. 181 and *Juniperus* p. 7.

Thymus vulgaris L. (garden thyme) C; see p. 182.

Tilia spp. (basswood) CN; pollen, AA: 109.

Tithymalopsis ipecacuanhae (L.) Small. CN; see *Euphorbia ipecacuanhae*.

Toxicodendron spp. N; see text p. 111.

T. quercifolium (Michx.) Greene (poison oak) N; see text p. 111.

T. radicans (L.) Kuntze (poison ivy) N; see text p. 111.

T. vernix (L.) Kuntze (poison ash, poison sumac) N; see text p. 111.

Tradescantia pallida Hunt synonym: *Setcreasea purpurea* Boom (purple queen) CN; see text p. 21.

Tragia spp. N; irritant hairs with a calcium oxalate tip (55); D: 109.

Trianthema portulacastrum L. (sea purslane) N; leaves, stems with soluble oxalates; roots used in India to induce abortion; I: 128.

Tribulus terrestris L. (puncture weed) CN; all parts with phylloerythrin (91); P: 142*, 165.

Trifolium spp. (clover) CN; some species with HCN (109).

T. hybridum L. (alslike clover) C; all parts, P: 142*, 165; D: 72, 142, 165; lotaustralin, linamarin; up to 350 mg HCN/100g of tissue in young leaves (145).

Trillium spp. (birthwort) CN; rhizomes, roots; small amounts medicinal; I: 90, 138, 142.

Tulipa spp. (tulip) C; bulb, stem, flower with tulipalin A, a phytoalexin that causes "tulip fingers" when the plant is bruised or damaged (55).

Turbina corymbosa (L.) Raf. synonym: *Rivea corymbosa* (L.) Hallier (Christmas pops) CN; seeds with alkaloids ergine, isoergine, chanoclavine and others (142); hallucinogenic, see *Ipomoea* (106); I: 106, 142.

Tussilago farfara L. (colt's foot) CN; used herbally; with pyrrolizidine alkaloids that are carcinogenic to rats and other laboratory animals (77).

Typha spp. (cattail) CN; airborne seeds and pollen, AA: 109.

Ulmus spp. (elm) CN; pollen dermatitis caused by oleoresins (109).

U. glabra Huds. CN; D: 106.

U. procera Salisb. (English elm) CN; D: 106.

Urtica spp. CN; pollen, AA: 109; stinging hairs (90, 109, 143).

U. chamaedryoides Pursh N; stinging hairs (106).

U. dioica L. (stinging nettle) CN; see text p. 41.

U. urens L. N; stinging hairs (106, 142).

Vanilla planifolia Andr. (Mexican vanilla) C; juice of vine, D: 142s; pollen, D: 128; may cause allergic reactions (108); grown in greenhouses (127).

Veratrum spp. (hellebore) N; leaves; D: 72.

V. parviflorum (Michx.) S. Wats. N; see *Melanthium parviflorum*.

V. virginicum (L.) Ait. CN; see *Melanthium virginicum*.

V. viride Aiton (false hellebore) CN; see text p. 30.

V. woodii Robbins N; see *Melanthium woodii*.

Verbascum spp. (mullein) CN; pollen, AA: 109; D: 109.

Vernonia noveboracensis (L.) Michx. CN; leaves; I: 101, 165.

Veronicastrum virginicum (L.) Farway (Culver's root) CN; roots medicinal and may be toxic in excess amounts (113).

Vicia faba L. (fava bean) C; see text p. 88.

V. sativa L. (vetch) C; see text p. 88.

Vigna sinensis (L.) Savi. C; see *V. unguiculata*.

V. unguiculata (L.) Walpers. synonym: *Vigna sinensis* (L.) Savi. (cowpea); C; raw seeds are slightly cyanogenic with 2.1 mg HCN/100 g tissue (145).

Vinca major L. (greater periwinkle) C; all parts with the alkaloids vincamajinine, reserpinine and others; used to induce abortion (142).

V. minor L. (common periwinkle) C; alkaloids isovincamine, minorine and others; I: 142s.

V. rosea L. CN; see text p. 141.

Viola odorata L. (English violet) C; medicinal; large doses may cause severe gastroenteritis, nervousness, respiratory and circulatory depression (109).

V. tricolor L. (pansy) CN; all parts medicinal, externally; excessive doses or extended use may cause skin problems (113).

Vitex trifoliata L.f. C; see notes p. 182.

Wisteria spp. CN; see text p. 90.

W. floribunda (Willd.) DC. (Japanese wisteria) C; wisterin; I: 142.

W. frutescens (L.) Poir. (American wisteria) C; as above; I: 142.

W. machrostachya (Torr. & Gray) Nutt. ex Robb & Fern. (Kentucky wisteria) C; as above; I: 142.

W. sinensis Sweet (Chinese wisteria) C; as above; I: 142.

Xanthium spp. (cocklebur) N; contact dermatitis caused by lactones (109); pollen dermatitis, oleoresins (109); seeds and young seedlings with hydroquinone are toxic to livestock, especially pigs (142).

X. spinosum L. (spiny cocklebur) N; D: 106.

X. strumarium L. (common cocklebur) N; D: 106.

Xanthosoma spp. (malanga, elephant's ear) C; sap; gastroenteric irritation, mostly mouth and throat; calcium oxalate crystals (106); I: 106, 127, 128; D: 101, 106, 109, 128.

X. jacquinii Schott (giant coco yam) C; I: 127.

X. violaceum Schott. C; see *Xanthosoma nigrum*.

X. nigrum Mansf. synonym: *X. violaceum* Schott (purple-stemmed elephant's ear) C; as in *Xanthosoma spp.*; I: 127, 128, 142; D: 128, 142.

Ximenia americana L. (purge nut) N; raw fruit kernel, toxic in large amounts (142); various parts medicinal (127).

Yucca spp. (Spanish bayonet) C; roots; I: 138d, 165sd.

Zamia spp. (coontie, Florida arrowroot) CN; see text p. 5.

Z. floridana DC. CN; unprepared root; I: 127, 128.

Z. integrifolia Aiton CN; as above; I: 127, 142d.

Zantedeschia spp. C; all parts with calcium oxalate crystals and unknowns; I, D: 142s.

Z. aethiopica (L.) Sprengel (common calla) C; all parts as above; gastroenteric irritant, primarily mouth and throat (106); I, D: 106, 142.

Zea mays L. (corn) C; pollen, AA: 109, 142s.

Zephyranthes atamasco (L.) Herb. (zephyr lily) CN; all parts especially bulbs; I: 90*, 101*, 138*, 142*, 143, 165.

Z. brazosensis Traub synonym: *Cooperia drummondii* Herb. C; bulbs with possibly lycorine and others; I: 142s?.

Zigandenus spp. (death camas) N; flowers, mostly bulbs with zygacine (109); I: 72, 143.

Z. densus (Desr.) Fernald N; see text p. 32.

Z. glaberrimus Michx. N; see text p. 32.

Z. muscaetoxicum (Walter) Zimm. CN; see *Amianthium*.

Z. nuttallii (Gray) S. Wats. CN; bulbs, leaves, flowers; I: 49sd.

Zingiber officinale Roscoe (ginger) C; allergic contact dermatitis during food use and preparation (165).

GLOSSARY

achene: a dry, indehiscent, 1-seeded fruit.
acuminate: tapering to a long point.
alternate leaves: those arising singly along the stem; not in pairs or whorls.
annual plant: one that completes its life cycle and dies in one year.
anther: the sac-like part of a stamen that holds the pollen.
anthraquinone: a yellowish glycoside resulting from the oxidation of anthracene.
anthelminthic oil: an oil taken from plants such as pigweed (*Chenopodium spp.*); it is used to treat intestinal worms.
aril: a fleshy appendage around a seed as in *Taxus*.
ascites: an accumulation of fluid in the abdominal cavity; a kind of dropsy.
ataxia: an inability to coordinate bodily movements.
auricle: an ear-shaped appendage or lobe.
awn: a slender bristle at the end or back of a structure.
axil: the angle formed between the upper side of a leaf stalk and the main stem.
basal rosette: a crowded cluster of leaves appearing to grow directly out of the ground.
berry: a simple, fleshy fruit with few to many seeds.
biennial plant: one that completes its life cycle and dies within 2 years.
bipinnate: doubly pinnate.
bisexual: flowers with both stamens and pistils.
blade: the flat portion of a leaf, petal or sepal.
bract: small leaf-like structures often at the base of a flower or flower cluster.
branchlet: last season's growth on a stem or branch.
bristle: a strong stiff hair.
bulb: usually a subterranean bud with fleshy bracts or food-storing leaves.
caducous: falling early; as in caducous sepals.
calyx: the sepals of a flower collectively; the outer whorl of floral parts.
calyx tube: a tube formed by the fusion of sepals.
campanulate: bell-shaped.
capillary bristles: hair-like bristles.
capitate: head-like.
capsule: a dry, usually dehiscent fruit with more than 1 carpel.
carpel: a portion of an ovary.
catharsis: purging of the bowels.
catkin: a spike-like inflorescence with unisexual flowers, without petals, as in willow; ament.
cauline: related to the stem; cauline leaves arise from the stem.
caulis: the main stem of a plant.
circulatory: referring to the bloodstream.
colic: acute abdominal pains caused by abnormal bowel conditions.
coma: a state of deep and prolonged unconsciousness.
compound leaf: the blade is composed of more than 1 leaflet.
compound pistil: has 2 or more united carpels.
conjunctivitis: inflammation of the mucous membrane covering the inner eyelids and eyeballs.
convulsion: a violent, involuntary contraction or spasm.
cordate: heart-shaped.
corm: a solid bulb as in autumn crocus (*Colchium*).
corolla: the petals collectively; the second from outside whorl of floral parts.
corona: a crown of inner petal-like appendages as in daffodil (*Narcissus*).
corymb: a raceme with the lower flower stalks longer than those above, making a flat-topped cluster.
crenate: scalloped; shallowly toothed.

Glossary

cyanogenic glycoside: one that yields HCN on hydrolysis.
cyme: a broad, determinate inflorescence; the central flowers bloom first.
deciduous: dropping leaves annually or seasonally.
decompound: more than once compound or divided.
decumbent: stems resting on the ground but with tips ascending.
decurrent: attached to the stem and extending beyond the point of attachment.
dehiscent fruit: one that splits open at maturity.
dermatitis: skin inflammation; sometimes with welts and eruptions.
determinate: an inflorescence with a single terminal flower opening before the lower ones as in a cyme.
diadelphous: stamens with filaments united into 2 often unequal fascicles.
diaphoretic: a medicine or treatment that induces or increases perspiration.
dioecious: said of plants with staminate and pistillate flowers on different individuals.
disc flowers: small tubular flowers in the middle of a floral head as in the aster family.
distichous: 2-ranked, on opposite sides of a stem, in the same plane.
drupe: a fleshy fruit with 1 hard and stony fruit.
emesis: to vomit.
emetic: a medicine or substance that induces vomiting.
emmenagogue: an agent that promotes menstrual flow.
entire: describes a smooth margin without teeth, lobes or divisions.
enzyme: an organic substance produced in plants and animals that catalyzes or aids chemical reactions.
escaped: said of ornamental plants that successfully reproduce on their own in nature.
evergreen: a plant that bears leaves year round.
exserted: projecting out, beyond.
fascicle: a dense cluster or bundle.
filament: the slender stalk of a stamen.
floral tube: a tube consisting of united petals or sepals.
follicle: a dry, dehiscent, 1-carpelled fruit opening only along one side as in milkweed (*Asclepias*).
fruit: a ripened ovary with the attached parts.
gastroenteritis: inflammation of the stomach and intestines by bacteria, viruses or toxins.
glabrous: without hairs.
gland: a small structure that usually secretes oil or nectar.
glaucous: covered with a powdery whitish-gray substance.
hastate: arrow-shaped, with spreading basal lobes.
head: a crowded cluster of flowers on a receptacle as in the aster family.
hematuria: the presence of blood or blood cells in the urine.
hemolysis: the destruction of red blood cells.
hepatogenic: derived from or originating in the liver; a hepatogenic disease is one of the liver.
herb: a non-woody plant, dying at least to ground level seasonally.
hirsute: with coarse or stiff hairs.
husk: the dry outer cover of some fruits as in the walnut.
hybrid: offspring resulting from mating different kinds of plants or animals.
hydrolysis: a splitting of compounds into simpler ones by the chemical addition of water.
hypanthium: a cup-like base composed of the united bases of the calyx, corolla and stamens.
hypokalemia: a deficiency of potassium in the blood.
imbricate: overlapping as shingles do.
imperfect flower: one lacking either pistils or stamens.
indehiscent fruit: a dry fruit that does not split open at maturity.

Glossary

inflorescence: the arrangement of flowers on a plant.
inserted: attached to another part or organ; within; as opposed to exserted.
internodes: the part of a stem between successive nodes.
introduced: refers to a plant brought into a new area where it is not native.
irregular flower: one with different size petals; zygomorphic.
involucre: a whorl or circle of bracts surrounding a flower, flower cluster or head.
keel: a sharp ridge; in a pea flower, the lowest petals that are united.
lacrimation: tears; crying.
lactone: a glycoside that yields the aglycone coumarin.
lanceolate: lance-like; lanceolate leaves are widest below the middle.
leaflet: one of the separate divisions of the blade of a compound leaf.
legume: dry fruit developing from a single ovary, and opening along 2 sides; follicles open only along one side.
leukopenia: a condition in which the number of white blood cells in circulation is abnormally low.
linear: long, narrow, with parallel sides, as in the leaf blades of grasses.
lobed: margins indented, but not to the point of reaching the center or base.
locule: the cavity of an ovary or anther.
mace: the dry outer covering of the nutmeg fruit.
margin: the outer edge of a leaf blade.
midrib: the central vein of a pinnately veined leaf.
monadelphous: stamens with filaments united into a single tube.
monoecious: having both staminate and pistillate flowers on the same individual.
monotypic: of a single type; for example, a genus containing only one species.
naturalized: said of an introduced species that can survive on its own.
nerve: a simple or unbranched vein or rib.
neuropathy: a nervous disorder.
node: the part of a stem where leaves and axillary buds arise.
nut: indehiscent, hard, 1-seeded fruit larger than an achene or nutlet.
nutlct: a 1-seeded portion of a fruit that fragments at maturity.
ob-: a Latin prefix, usually meaning inversion.
obcordate: inverted heart-shaped, with the point being basal (attached to the stalk).
oblique: unequal sided.
oblong: longer than broad with more or less parallel sides.
obovate: inverted egg-shaped.
obsolete: not evident or only rudimentary.
obtuse: blunt or rounded at the tip.
ocrea: leaf bases that sheath the stem in members of the buckwheat family (Polygonaceae).
oliguria: reduced output of urine.
opposite leaves: occur in pairs along the stem.
orbicular: circular.
ovary: the swollen pistil base where seeds develop.
ovate: egg-shaped with the basal end broader.
ovule: immature seed in an ovary.
palmate leaf: has lobes radiating from a central point.
panicle: a branched raceme; each branch bears a raceme of flowers.
papillose: short, rounded, warty knobs.
pappus: bristles, scales or a crown on the achenes of the aster family; a modified calyx.
pedicel: stalk supporting a single flower.
peduncle: stalk supporting a flower cluster.
pemphigus: a condition characterized by the formation of water blisters on the skin.
pendulous: drooping or hanging down.

Glossary

perennial: duration of 3 or more years.
perfect flower: one with both stamens and pistil(s).
perianth: the calyx and corolla together.
pericarp: mature ovary wall.
persisting: remaining attached.
petal: one member of the corolla; usually flat, broad and colored.
petaloid: petal-like; a colored sepal.
petiole: leaf stalk; attaches the blade to the stem.
pharynx: the cavity between the mouth, larynx and esophagus.
pilose: with soft hairs.
pinna: a division of a pinnate leaf.
pinnate leaf: a compound leaf with leaflets on both sides of a common axis.
pinnule: a secondary pinna or leaflet.
pistil: female seed-bearing organ of the flower; the ovary, stigma and style collectively.
pistillate flower: female, flower, with 1 or more pistils, but no functioning stamens.
placenta: ovary tissue where ovules attach.
pod: a dry fruit opening at maturity, as in the bean family.
pollen: yellow, powdery, male sex cells produced by the stamens of a flower.
polygamous: bearing unisexual or bisexual flowers on the same plant or different individuals of the same species.
polyuria: excessive urination.
pome: fleshy fruit from an inferior ovary embedded in receptacle tissue, as an apple.
pubescent: hairy.
punctate: dotted with small holes or indentations.
purgative: a bowel emptying substance.
raceme: unbranched inflorescence of stalked flowers on a common axis.
raphides: sharp crystals, usually of calcium oxalate formed within plant cells; considered as metabolic waste.
ray flower: flowers around the edge of a composite head in the aster family.
receptacle: flower base where all parts are attached.
reflexed margin: one that is curved or bent down.
regular flower: one that has petals and sepals of equal size.
renal: referring to the kidneys.
reniform: kidney-shaped.
reticulate: with a network of veins or nerves.
retrorse: with spines or hairs pointing down or backwards.
rhizome: underground stem, with nodes.
rootstock: a rhizome.
rotund: plump.
salverform: describing a slender tube that expands into a wide opening.
scabrous: rough to the touch; covered with stiff, short hairs.
scale-like leaves: small, dry, appressed leaves or bracts.
scape: leafless flowering stem arising from the ground.
sepal: one division of a calyx; the outer whorl of floral parts.
serrate: saw-toothed.
sessile: without a stalk or stem.
shrub: a bush; a woody plant with several stems, usually less than 6 m tall.
silique: long, slender, many-seeded fruit in which the 2 valves that split from the bottom are separated by a false partition as in the mustard family (Brassicaceae).
simple leaf: one with an undivided blade.
somnolent: sleepy or drowsy.
spadix: dense spike of flowers, usually enclosed by a spathe as in the arum family (Araceae).
spasm: a sudden, involuntary muscle contraction.

Glossary

spathe: a bract or pair of bracts enclosing an inflorescence.
spike: unbranched inflorescence of sessile flowers.
stamen: male, pollen-bearing organ in flowers.
staminate flowers: male flower producing pollen; without pistils.
standard: upper petal of a pea flower; the banner or vexillum.
stigma: tip of the pistil where the pollen lands.
stipitate: with a stalk-like base.
stipule: a pair of small appendages at the base of a leaf.
stomititis: inflammation of the mouth.
stupor: a state in which the mind and senses are dulled.
style: narrow tube-like part of the pistil, connecting the ovary and stigma.
succulent: fleshy and thick, storing water; as in the leaves of cactus plants.
tachycardia: accelerated heartbeat.
tachypnea: accelerated and difficult breathing.
tannin: tannic acid; a yellowish, astringent substance from the bark of some trees and plants.
tendril: thread-like outgrowth adapted for twining; found on vines.
terete: round in cross-section.
terminal: occurring at the end or extremity of something.
thyrse: panicle-like inflorescence in which the main axis is indeterminate and side axes determinate.
tomentose: densely pubescent with matted wool.
trichome: simple or branched hair-like structure.
trifoliate: a compound leaf with 3 leaflets.
tuber: a short, thickened underground stem.
umbel: inflorescence in which pedicels and/or peduncles arise from a common point.
undulate: with a wavy edge.
unisexual: staminate or pistillate only.
urethra: the canal in which urine is discharged from the bladder.
utricle: small, bladdery, 1-seeded fruit.
valve (valvate): one of the pieces into which a capsule splits; the partially detached lid of an anther.
variegated: of different colors, in spots or streaks.
variety: a group within a species having characteristics of its own.
veins: strands of vascular conducting tissue in leaves, etc.
vesicant: a substance that causes blisters.
whorled: leaves in a circle around the stem.
wing: the lateral petal of a pea-like corolla.
woolly: with long and matted hairs.

LITERATURE CITED

1. Adolf, W. and Hecker, E. On the Active Principle of the Spurge Family, III. Skin Irritants and Cocarcinogenic Factors from the Caper Spurge. Z. krebsforsch. 84:325–44. 1975. (As cited by Kinghorn, 1983).

2. Adolf, W. and Hecker, E. On the Irritant and Cocarcinogenic Principles of *Hippomane mancinella*. Tetrahedron Lett. 1587–90. 1975.

3. Allen, J., Robertson, K., Johnson, W., and Carstens, C. Toxicological Effects of Monocrotaline and Its Metabolites. (in Cheeke, 1979).

4. Altman, H. Poisonous Plants and Animals. London: Chatto & Windus Ltd.; 1980.

5. Angier, B. Field Guide to Medicinal Plants. Harrisburg, PA: Stackpole Books; 1978.

6. Anonymous. The Dangers of Eating Rhubarb Leaves. Sci. Am. 117:82. 1917.

7. ———. Toxicity Studies of Arizona Ornamental Plants. Ariz. Med. 15:512. 1958.

8. ———. Rhododendron Ingestion. Bulletin of the National Clearinghouse for Poison Control Centers. DHEW, FDA, Bureau of Drugs. 1973.

9. Ansford, A. and Morris, H. Oleander Poisoning. Toxicon. (Suppl. 3) pp.15–16. 1983.

10. Arena, J. Poisoning: Toxicology, Symptoms, Treatment. 3rd edition. Springfield, IL: Charles C. Thomas; 1974.

11. Arnold, H. L. Poisonous Plants of Hawaii. Rutland, VT: C. Tuttle, Co.; 1968.

12. Ayres, S., Jr. and Ayres, S., III. Philodendron as a Cause of Contact Dermatitis. AMA Arch. Derm. 78:330–3. 1958.

13. Bacon, A. An Experiment with the Fruit of Red Baneberry. Rhodora. 5:77. 1903.

14. Baer, H. Allergic Contact Dermatitis from Plants. (in Keeler and Tu, 1983. pp. 421–42).

15. Balint, G. Ricin: The Toxic Protein of Castor Oil Seeds. Toxicology. 2:77–102. 1974.

16. Balthrop, E. Tung Nut Poisoning. South. Med. J. 45:864. 1952.

17. Balucani, M. and Zellers, D. Podophyllum Resin Poisoning with Complete Recovery. JAMA. 189(8):639. 1964.

18. Bandelin, F. and Malesh, W. Alkaloids of *Chelidonium majus* L. Leaves and Stem. J. Amer. Pharm. Assoc. 45:702. 1956.

19. Batson, W. T. Professor Emeritus, University of South Carolina, Columbia. Personal communication, 7-19-84.

20. Becker, L. and Skipworth, G. Ginkgo Tree Dermatitis. Stomatitis and Proctitis. JAMA. 231:1160–3. 1975.

21. Berenblum, I. The Carcinogenic Action of *Croton* Resin. Cancer Res. pp. 144–8. 1941.

22. Bianchini, F., Corbetta, F. and Pistoia, M. Health Plants of the World. New York: Newsweek Books; 1979.

Literature Cited

23. Blaw, M., Adkisson, M., Garriott, J. and Tindall, T. Poisoning with Carolina Jessamine, *Gelsemium sempervirens* (L.) Ait. J. Pediatr. 94(6):998–1001. 1979.
24. Blumstein, G. Buckwheat Sensitivity. J. Allergy. 7:74. 1935.
25. Brondegaard, V. Contraceptive Plant Drugs. Planta Med. 23:167–72. 1973.
26. Brown, A. and Brown, F. Mango Dermatitis. J. Allergy. 12:310. 1940.
27. Brown, P., VonEuw, J., Reichstein, T., Stockel, K. and Watson, T. Cardenolides of *Asclepias syriaca* L., Probable Structure of Syrioside and Syriobioside. Helv. Chim. Acta. 62:412–41. 1979. (as cited by Seiber et al., 1983).
28. Buere, P. A Coffee Substitute, *Cassia occidentalis,* That Is Toxic Before Roasting. J. Pharm. Chim. 9(2):321–4. 1942.
29. Bruschweiler, F., Stockel, K., and Reichstein, T. *Calotropis* Glycosides, Presumed Partial Structure. Helv. Chim. Acta. 52:2276–2303. 1969. (as cited by Seiber et al., 1983).
30. Caldron-Gonzalez, R. and Rizzi-Hernandez, H. Buckthorn Neuropathy. New Engl. J. Med. 277:69. 1967.
31. Cameron, K. Death Camas Poisoning. North. Med. 51:682. 1952.
32. Carlton, B., Tupts, E. and Girard, D. Water Hemlock Poisoning Complicated by Rhabdomyolysis and Renal Failure. Clin. Toxicol. 14(1):87–92. 1979.
33. Cheeke, P. R., Editor. Symposium on Pyrrolizidine (*Senecio*) Alkaloids: Toxicity, Metabolism and Poisonous Plant Control Measures. Oregon State University, Corvallis: Nutrition Research Institute; 1979.
34. Chen, K. and Chen, A. The Action of Crystalline Thevatin, a Cardiac glucoside of *Thevetia neriifolia*. J. Pharmacol. Exp. Ther. 51:23. 1934.
35. Chesnut, V. Principal Poisonous Plants of the United States. USDA, Division of Botany. Bulletin 20. 1898.
36. Chin. Med. J. (Engl.) Gossypol—A New Anti-Fertility Agent for Males. 4(6):417–28. 1978.
37. Cohen, S. Suicide Following Morning Glory Ingestion. Am. J. Psychiatry. 120:1024–5. 1964.
38. Coursey, D. Cassava as Food: Toxicity and Technology. in: Chronic Cassava Toxicity. Nestel, B., and MacIntyre, R., Editors. Ottawa: Intl. Develop. Research Center; pp. 27–36. 1973.
39. Curtis, D. and Johnston, G. Convulsant Alkaloids. (in: Simpson and Curtis, 1974. pp. 207–48).
40. DeBoer, J. The Death of Socrates. A Historical and Experimental Study on the Actions of Coniine and *Conium maculatum*. Arch. Internat. Pharmacodyn. Ther. 83:473–90. 1950.
41. Deinzer, M., Thomson, P., Burgett, D. and Isaacson, D. Pyrrolizidine Alkaloids: Their Occurrence in Honey from Tansy Ragwort (*Senecio jacobea* L.). Science (Washington, DC). 195:497–9. 1977.
42. Deinzer, M., Thomson, P., Griffin, D. and Burgett, D. The Analysis of Pyrrolizidine Alkaloids in Agricultural Food Stuffs. (in Cheeke, 1979).
43. DerMardosian, A. Poisonous Plants in and Around the Home. Am. J. Pharm. Educ. 30:115–40. 1966.
44. DerMarderosian, A., Giller, F. and Roia, F., Jr. Phytochemical and Toxicological Screening of Household Plants Potentially Toxic to Humans. J. Toxicol. Environ. Health. 1:939–53. 1976.

Literature Cited

45. Dickinson, J., Cooke, M., King, P. and Mohamed, P. Milk Transfer of Pyrrolizidine Alkaloids in Cattle. J. Am. Vet. Med. Assoc. 169:1192–6. 1976.

46. Drach, G. and Maloney, W. Toxicity of the Common House Plant *Dieffenbachia*. JAMA. 184:1047. 1963.

47. Drever, J. and Hunter, J. Giant Hogweed Dermatitis. Scott. Med. J. 15:315. 1970.

48. Durant, M. Who Named the Daisy? Who Named the Rose? New York: Dodd, Mead & Co.; 1976.

49. Ellis, M. Dangerous Plant, Snakes, Arthropods and Marine Life of Texas. Public Health Service, U.S. Dept. of HEW; 1975.

50. Emery, Z. Report of Thirty-Two cases of Poisoning by Locust Bark. N.Y. Med. J. 45:92. 1887.

51. Ellwood, M. and Robb, G. Self-Poisoning with Colchicine. Postgraduate Med. J. 47:129. 1971.

52. Erickson, J. and Brown, J., Jr. A Study of the Toxic Properties of Tung Nuts. J. Pharmacol. Exp. Ther. 74:114. 1942.

53. Evans, I., Widdop, B., Jones, R., Barber, G., Leach, H., Jones, D. and Mainwaring-Burton, R. The Possible Human Hazard of the Naturally Occurring Bracken Carcinogen. Biochem. J. 124:28p–29p. 1971.

54. Evans, I., Jones, R. and Mainwaring-Burton, R. Passage of Bracken Fern Toxicity into Milk. Nature (London). 273:107–8. 1972.

55. Evans, F. and Schmidt, R. Plants and Plant Products That Induce Contact Dermatitis. Planta Med. 38(4):289–316. 1980.

56. Franke, F. and Thomas, J. The Treatment of Acute Nicotine Poisoning. JAMA. 106(7):509–12. 1936.

57. French, C. Pokeroot Poisoning. N.Y. Med. J. 72:653. 1900.

58. Frohne, D. and Pribilla, O. Todliche Vergiftung mit. *Taxus baccata*. Arch. Toxikol. 21:150. 1965. (as cited by Lampe and Fagerstrom, 1968).

59. Furstenburger, G. and Hecker, E. New Highly Irritant Euphorbia Factors from Latex of *Euphorbia tirucalli* L. Experientia. 33:986–8. 1977.

60. Gellerman, J., Anderson, W. and Schenk, H. 6-(pentadec-8-enyl)-2, 4 dihydroxybenzoic Acid from Seeds of *Ginkgo biloba*. Phytochemistry. 15:1959–61. 1976.

61. Giese, A. Photosensitivity by Natural Pigments. in: Photophysiology. Vol. VI. A. C. Giese, Editor. New York: Academic Press; pp. 77–129. 1971.

62. Gompertz, L. Poisoning with Water Hemlock (*Cicuta maculata*). JAMA. 87:1277. 1926.

63. Gooneratne, B. Massive Generalized Alopecia After Poisoning by *Gloriosa superba*. Br. Med. J. 1:1023. 1966.

64. Grant, L. A Brief Study of the Corn Cockle as an Allergen. J. Allergy. 8(5):506. 1937

65. Gross, H., Jones, W., Cook, E. and Boone, C. Carcinogenicity of Oil of Calamus. Proc. Am. Assoc. Cancer Res. 8:24. 1967.

66. Gross, M., Baer, H. and Fales, H. Urushiols of Poisonous Anacardiaceae. Phytochemistry. 14:2261–6. 1975.

67. Guenther, E. and Althausen, D. The Essential Oils. Vol. 2. The Constituents of Volatile Oils. Princeton, NJ: Van Nostrand; 1949.

Literature Cited

68. Gunby, P. Plant Known for Centuries Still Causes Problems Today. JAMA. 241(21):2246–7. 1979.
69. Hakim, S., Mikovic, V. and Walker, J. Experimental Transmission of Sanguinarine in Milk: Detection of a Metabolic Product. Nature (London). 189:201–4. 1961.
70. Hansen, A. Two Fatal Cases of Potato Poisoning. Science (Washington, DC). 61:340. 1925.
71. Harborne, J. Toxins of Plant-Fungal Interactions. (in: Keeler and Tu, 1983. pp. 743–84).
72. Hardin, J. and Arena, J. Human Poisoning from Native and Introduced Plants. 2nd edition. Durham, NC: Duke University Press; 1974.
73. Hart, M. Hazards to Health. Jequirity-Bean Poisoning. New Engl. J. Med. 268:885. 1963.
74. Hesse, G., Heuser, L. and Reichneder, F. African Arrow Poisons. IV. Relationships Between the most Important Poisons of *Calotropis procera*. Liebigs Ann. Chem. 566:136–9. 1950. (as cited by Seiber et al., 1983).
75. Hickey, M. Investigation of the Chemical Constituents of Brazilian Sassafras Oil. J. Org. Chem. 13:443–6. 1948.
76. Hikino, H., Ohizumi, Y., Konno, C., Hashimoto, K. and Wakasa, M. Subchronic Toxicity of Ericaceous Toxins and Rhododendron Leaves. Chem. Pharm. Bull. (Tokyo). 24(4):874–9. 1979.
77. Hirono, I., Mori, H. and Culvenor, C. Carcinogenic Activity of Coltsfoot, *Tussilago farafara* L. Gann. 67:125–9. 1976.
78. Hirono, I., Mori, H., and Haga, M. Carcinogenic Activity of *Symphytum officinale*. J. Natl. Cancer Inst. 61:865–9. 1978.
79. Hjorth, N., Frcgert, S. and Schildnecht, H. Cross-Sensitization Between Synthetic Primin and Related Quinones. Acta Derm. venereol. (Stockh). 49:552–5. 1969.
80. Ho, R. Acute Poisoning from the Ingestion of Seeds of *Jatropha curcas*. Report of Five Cases. Hawaii Med. J. 19:421. 1960.
81. Hoffmann, D., Hecht, S., Ornaf, R. and Wynder, E. N'-Nitrosonornicotine in Tobacco. Science (Washington, DC). 186:265–76. 1974.
82. Homberger, F., Kelley, T., Jr., Friedler, G. and Russfield, A. Toxic and Possible Carcinogenic Effects of 4-allyl-1, 2-methylenedioxybenzene (Safrole) in Rats on Deficient Diets. Med. Exp. 4:1–11. 1961.
83. Hooper, P. Cycad Poisoning. (in: Keeler and Tu, 1983. pp. 463–72).
84. Hosel, W. and Nahrstedt, A. Specifische Glucosidasen fur das Cyanglucosid Triglochin: Reinigung und Charakterisierung von β-Glucosidasen aus *Alocasia macrorrhiza* Schott. Hoppe-Seylers Z. Physiol. Chem. 356:1265. 1975. (as cited by Poulton, 1983).
85. Hutton, J. Favism. An Unusually Observed Type of Hemolytic Anemia. JAMA. 109(20):1618–20. 1937.
86. Huxtable, R. Herbal Teas and Pyrrolidine Alkaloids. (in: Cheeke, 1979. pp. 87–94).
87. Hylin, J. Toxic Peptides and Amino Acids in Foods and Feeds. J. Agric. Food Chem. 17:492–6. 1969.
88. Ivie, G. and Witzel, D. Sesquiterpene Lactones: Structure, Biological Action and Toxicological Significance. (in: Keeler and Tu, 1983. pp. 543–84).

Literature Cited

89. Jacobziner, H. Accidental Chemical Poisonings (Jack-in-the-Pulpit). N.Y. State J. Med. 62:3130. 1962.
90. James, W. Know Your Poisonous Plants. Heraldsburg, CA: Naturegraph Publishers; 1973.
91. Johnson, A. Photosensitizing Toxins from Plants and Their Biologic Effects. (in: Keeler and Tu, 1983. pp. 345–59).
92. Johnston, G. Neurotoxic Amino Acids. (in: Simpson and Curtis, 1974. pp. 179–205).
93. Jordan, M. A Guide to Wild Plants. London: Millington Books, Ltd.; 1976.
94. Joshi, B. and Rane, D. The Structure of Diosbulbine. Indian J. Chem. 7:452–6. 1969.
95. Kean, B. Death Due to Akee Poisoning in Panama. Am. J. Trop. Med. 23(3):339. 1943.
96. Keeler, R. Toxins and Teratogens of the Solanaceae and Liliaceae. (in: Kinghorn, 1979. pp. 59–82).
97. ———. Naturally Occurring Teratogens from Plants. (in: Keeler and Tu, 1983. pp. 161 200).
98. Keeler, R. and Tu, A. (Editors). Handbook of Natural Toxins. Vol. I. Plant and Fungal Toxins. New York: Marcel, Dekker Inc.; 1983.
99. Kinghorn, A. (Editor). Toxic Plants. New York: Columbia University Press, 1979.
100. Kinghorn, A. Carcinogenic and Cocarcinogenic Toxins from Plants. (in: Keeler and Tu, 1983. pp. 239–98).
101. Kingsbury, J. Poisonous Plants of the United States and Canada. Englewood Cliffs, NJ: Prentice-Hall, Inc.; 1964.
102. Krochmal, A. and Krochmal, C. A Guide to the Medicinal Plants of the United States. New York: Quadrangle; New York Times Book Co.; 1973.
103. Krutch, J. Herbal. Boston: David Godine, Publ. 1976.
104. Kupchan, S. and Baxter, R. Mezerein: Anti-Leukemic Principle Derived from *Daphne mezereum* L. Science (Washington, DC). 187:652–3. 1975.
105. Lain, E. Dermatitis Due to *Lycopersicon esculentum* (tomato plant). JAMA. 71:114. 1918.
106. Lampe, K. and Fagerstrom, R. Plant Toxicity and Dermatitis: A Manual for Physicians. Baltimore: Williams and Wilkins; 1968.
107. Lancaster, A. *Clematis* Dermatitis. South. Med. J. 30(2):207. 1937.
108. Leung, A. Encyclopedia of Common Natural Ingredients Used in Food, Drugs and Cosmetics. New York: Wiley & Sons; 1980.
109. Lewis, W. and Elvin-Lewis, M. Medical Botany. Plants Affecting Man's Health. New York: Wiley & Sons; 1977.
110. ———. Contributions of Herbology to Modern Medicine and Dentistry. (in: Keeler and Tu, 1983. pp. 786–816).
111. Lindenbaum, S. Case Report: Pollinosis Due to *Ricinus communis* or Castor Bean Plant. Ann. Allergy. 24:23. 1966.
112. Los Angeles Times. Section B–8. 6–10–1962.
113. Lust, J. The Herb Book. New York: Benedict Lust Publications; 1974.
114. Matsumoto, H. and Strong, F. The Occurrence of Methyl-azoxymethanol in *Cycas circinalis* L. Arch. Biochem. Biophys. 101:299–310. 1963.

Literature Cited

115. Maugh, T., II. Research News: Male "Pill" Blocks Sperm Enzyme. Science (Washington, DC). 212:314. 1981

116. Mercatante, A. The Magic Garden: The Myth and Folklore of Flowers, Plants, Trees and Herbs. New York: Harper & Row; 1976.

117. Metts, B. Director, Palmetto Poison Control Center, University of South Carolina, Columbia. personal communication, 5–14–84.

118. Mikolich, J., Paulson, G. and Cross, C. Acute Anticholinergic Syndrome Due to Jimson Seed Ingestion. Clinical and Laboratory Observations in Six Cases. Ann. Intern. Med. 83:321–5. 1975.

119. Minors, E. Five Cases of Belladonna Poisoning. Br. Med. J. p. 518. Sept. 1948.

120. Mitchell, J. and Jordan, W. Allergic Contact Dermatitis from the Radish, *Raphanus sativus*. Br. J. Dermatol. 91:183–9. 1974.

121. Mitchell, R. *Laburnum* Poisoning in Children. Lancet. 261:57. 1951.

122. Montgomery, R. The Medical Significance of Cyanogen in Food Stuffs. Am. J. Clin. Nutr. 17:103–13. 1965.

123. Moore, H. Mistletoe Poisoning. J. S.C. Med. Assoc. 59(8):269. 1963.

124. Morimoto, H., Kawamatsu, Y. and Sugihara, H. Sterische Stuktur der Giftstoffe aus dem fruchtfleisch von *Ginkgo biloba*. Chem. Pharm. Bull. (Tokyo). 16:2282–6. 1968. (as cited by Baer, 1983).

125. Morton, J. Cajeput Tree, a Boon and an Affliction. Econ. Bot. 20(1):31. 1966.

126. ———. Folk Remedies of the Low Country. Miami: Seemann Publ. Inc.; 1974.

127. ———. 500 Plants of South Florida. Miami: Seemann Publ. Inc.; 1974.

128. ———. Plants Poisonous to People in Florida and Other Warm Areas. 2nd edition. Published by the Author. 1982.

129. Muenscher, W. Poisonous Plants of the United States. New York: Collier Books; 1975.

130. Nagarajan, V., Mohan, V. and Gopalan, C. Further Studies on the Toxic Factor in *Lathyrus sativus*. Potentiation of a Toxic Fraction from the Seed by Some Amino Acids. Indian J. Biochem. 3:130. 1966.

131. Nelson, D. Accidental Poisoning by *Veratrum japonicum*. JAMA. 156:133. 1954.

132. Nishida, K., Kobayashi, A. and Nagahama, T. Cycasin, a Toxic Glycoside of *Cycas revoluta*. I. Isolation and Structure of Cycasin. Bull. Agric. Chem. Soc. Jpn. 19:77–84. 1955. (through: Chem. Abstr. 50:13756g. 1956).

133. Nishida, K., Kobayashi, A. and Nagahama, T. Some New Azoxy Glycosides of *Cycas revoluta*. I. Neocycasin A. Bull. Agric. Chem. Soc. Jpn. 23:460–4. 1959 (through: Chem. Abstr. 54:3609d. 1960).

134. North, P. Poisonous Plants and Fungi. London: Blandford Press Ltd.; 1967.

135. Oakes, A. and Butcher, J. Poisonous and Injurious Plants of the U.S. Virgin Islands. USDA ARS Misc. Publication #882. 1962.

136. Oberst, B. and McEntyre, R. Acute Nicotine Poisoning. J. Pediatr. 2:338–40. 1953.

137. O'Brien, R. Atropine. (in: Simpson and Curtis, 1974. pp. 157–78).

138. O'Leary, S. Poisoning in Man from Eating Poisonous Plants. Arch. Environ. Health. 9:216–42. 1964.

Literature Cited

139. Olsnes, S. and Pihl, A. Abrin, Ricin and Their Associated Agglutins. in Receptors and Recognition. Series B. Vol. I. The Specificity and Action of Animal, Bacterial and Plant Toxins. P. Cuatrecasus, (Editor). New York: Wiley & Sons; pp. 129–74. 1977.

140. Osterloh, J., Herold, S. and Pond, S. Oleander Interference in the Digoxin Radioimmunoassay in a Fatal Ingestion. JAMA. 247(11):1596–7. 1982.

141. Panter, K., Keeler, R., Buck, W. and Shupe, J. Toxicity and Teratogenicity of *Conium maculatum* in Swine. Toxicon. (Suppl. 3) pp. 333–6. 1983.

142. Perkins, K. and Payne, W. Guide to the Poisonous and Irritant Plants of Florida. University of Florida, Gainesville: Circular 441; Coop. Ext. Service; Inst. Food and Agric. Sci.; 1978.

143. Peterson, L. A Field Guide to Edible Wild Plants of Eastern and Central America. Boston: Houghton Mifflin Co.; 1978.

144. Polhamus, L., Hill, H. and Elder, J. The Rubber Content of Two Species of *Cryptostegia* and of an Interspecific Hybrid in Florida. USDA. Technical Bulletin 457. 1934.

145. Poulton, J. Cyanogenic Compounds in Plants and Their Toxic Effects. (in: Keeler and Tu, 1983. pp. 117–57).

146. Rao, S., Adiga, P. and Sarma, P. The Isolation and Characterization of β-N-oxalyl-L-α,β-diaminoproprionic Acid: A Neurotoxin from the Seeds of *Lathyrus sativus*. Biochemistry. 3:432. 1964.

147. Rascoff, H. and Wasser, S. Poisoning in a Child Simulating Diabetic Coma. JAMA. 152:1134. 1953.

148. Ressler, C. Isolation and Identification from Common Vetch of the Neurotoxin β-cyano-L-alanine, a Possible Factor in Neurolathyrism. J. Biol. Chem. 237:733. 1962.

149. Riggs, N. Glucosylvoazoxymethane, a Constituent of the Seeds of *Cycas circinalis* L. Chem. Ind. (London). P. 926. 1956.

150. Robeson, P. Water Hemlock Poisoning. Lancet. 2:1274. 1965.

151. Rubino, M. Cyanide Poisoning from Apricot Seeds. JAMA. 241(4):359. 1979.

152. Rukmini, C. Isolation and Purification of a New Toxic Factor from *Lathyrus sativus*. Indian J. Biochem. 5:182. 1968.

153. Sakata, K., Kawazu, K., Mitsui, T. and Masaki, N. Structure and Stereochemistry of Huratoxin, a Piscicidal Constituent of *Hura crepitans*. Tetrahedron Lett. pp. 1141–4. 1971.

154. Santos. Philippine J. Sc. 46:257. 1931. as cited by: Swanson, E. and Chen, K. The Pharmacological Action of Coriamyrtin. J. Pharmacol. Exp. Ther. 57:410. 1936.

155. Sarma, P. and Padamanaban, G. Lathyrogens. in: Toxic Constituents of Plant Foodstuffs. I. Liener, (Editor). New York: Academic Press; pp. 267–91. 1969.

156. Schmeltz, I. and Hoffmann, D. Nitrogen-Containing Compounds in Tobacco and Tobacco Smoke. Chem. Rev. 77: 295–311. 1977.

157. Schweitzer, H. Todliche Saponinvergiftung. durch Genuss von. Rosskastanien. Med. Klin. 47:683. 1952. (as cited by: Lampe and Fagerstrom, 1968).

158. Seiber, J., Lee, S. and Benson, J. Cardiac Glycosides (Cardenolides) in Species of *Asclepias* (Asclepiadaceae). (in: Keeler and Tu, 1983. pp. 43–83).

Literature Cited

159. Silverman, M. A City Herbal. A Guide to the Lore, Legend and Usefulness of 34 Plants That Grow Wild in the City. New York: Alfred A. Knopf; 1977.
160. Simmons, F. Jimsonweed Mydriasis in Farmers. Am. J. of Ophthalmol. 44:109. 1957.
161. Simpson, L. and Curtis, D., (Editors). Neuropoisons: Their Pathophysiological Actions. Vol. 2. Poisons of Plant Origin. New York: Plenum Press; 1974.
162. Singh, B. and Rastogi, R. Chemical Investigation of *Asclepias curassivica* Linn. Indian J. Chem. 7:1105–10. 1969.
163. Swan, G. An Introduction to the Alkaloids. New York: Wiley & Sons; 1967.
164. Tallqvist, H. and Vaananen, I. Death of a Child from Oxalic Acid Poisoning Due to Eating Rhubarb Leaves. Ann. Paediatr. Fenn. 6:744. 1960.
165. Tampion, J. Dangerous Plants. New York: Universe Book Co.; 1977.
166. Thomson, R. Naturally Occurring Quinones. 2nd edition. London: Academic Press; 1971.
167. Tsai, W. and Ling, K. Toxic Action of Mimosine. I. Toxicon. 9:241–7. 1971.
168. Van Doren, M., (Editor). Travels of William Bartram. New York: Dover Publ.; 1955. originally published in Philadelphia: 8–26–1781.
169. Viehoever, A. Edible and Poisonous Beans of the Lima Type (*Phaseolus lunatus* L.). Thai Sci. Bull. 2:1–99. 1940. (as cited by: Poulton, 1983).
170. Waller, D., Zaneveld, L. and Fong, H. In vitro Spermicidal Activity of Gossypol. Contraception. 22:183–7. 1980.
171. Watt, J. and Breyer-Brandwijk, M. Medicinal and Poisonous Plants of Southern and Eastern Africa. 2nd edition. Edinburgh and London: E. & S. Livingstone; 1962.
172. Weil, A. Nutmeg as a Narcotic. Econ. Bot. 19:194–217. 1965.
173. Wen, J. China Invents Male Birth Control Pill. Am. J. Chin. Med. 8:195–7. 1980.
174. West, E. Poisonous Plants Around the Home. University of Florida Agric. Exptl. Station. Circular # S–100. August, 1957.
175. Whiting, M. Toxicity of Cycads. Econ. Bot. 17:271–302. 1963.
176. Wiswell, O., Irwin, I., Guba, E., Rackemann, F. and Nerii, L. Contact Dermatitis of Celery Farmers. J. Allergy. 19:396. 1948.
177. Wolfson, S. and Solomons, I. Poisoning by Fruit of *Lantana camara*. Am. J. Dis. Child. 107:173. 1964.

GENERAL REFERENCES

Bailey, L. H. Hortus Third. New York: Macmillan Co.; 1976.
Batson, W. T. Genera of the Eastern Plants. New York: Wiley & Sons; 1977.
Cardenas, J., Reyes, C. and Doll, J. Tropical Weeds, Vol. I. Bogota, Colombia: Instituto Columbiano Agropecuaria; 1972.
Elias, T. S. The Complete Trees of North America. New York: VanNostrand and Reinhold Co.; 1980.
Fernald, M. L. Gray's Manual of Botany. 8th edition. New York: American Book Co.; 1950.
Hardin, J. Poisonous Plants of North Carolina. Agricultural Experiment Station. N.C. State University. Raleigh. Bulletin 414. 1962.
Kingsbury, J. Deadly Harvest (A Guide to Common Poisonous Plants). New York: Holt, Rinehart and Winston; 1965.
Lampe, K. Common Poisonous and Injurious Plants. U.S. Dept. HHS., Public Health Service. FDA, Bureau of Drugs, Division of Poison Control. Publication #: (FDA)81–7006.
Long, R. and Lakela, O. A Flora of Tropical Florida. A Manual of the Seed Plants and Ferns of Southern Peninsular Florida. Miami: Banyan Books; 1976.
Martin, F. and Ruberte, R. Edible Leaves of the Tropics. Published jointly by the Agency for Intl. Development, Department of State, and the Agricultural Research Service, USDA. 1975.
Mitchell, J. and Rook, A. Botanical Dermatology: Plants Injurious to the Skin. Vancouver, B.C.: Greenhouse Ltd.; 1979.
Radford, A., Ahles, H. and Bell, C. Manual of the Vascular Flora of the Carolinas. Chapel Hill: University of N.C. Press; 1968.
Terrell, E. A Checklist of Names for 3,000 Vascular Plants of Economic Importance. USDA ARS Handbook # 505. 1977.
USDA SCS. National List of Scientific Names. Publication # TP–159. 2 vol. 1982.

INDEX OF COMMON NAMES

adonis, spring, 173
agave, sisal, 184
agrimony, 184
akee, 117, 187
alder, 184
allamanda, pineland, 185; purple, 144, 190; yellow, 140
almond, bitter, 76, 201; sweet, 76, 201
allspice, Carolina, 188
amaryllis, 34, 185
ammi, greater, 185
anise, Japanese, 178, 195
apple, 74, 198; balsam, 165, 190, 198; custard, 185; hairy thorn, 192; pond, 185; sugar, 185; thorn, 68
apple-of-Peru, 199
apricot, 75, 201
aralia, 127, 201; geranium-leaf, 201
arbor-vitae, American, 181, 205
arnica, European, 173, 186
arrowroot, 5, 207
artichoke, 191
arum, arrow, 200; water, 187
ash, 194; mountain, 204; poison, 111, 205
asparagus, 22, 186
avens, water, 194
azalea, 134

bagpod, 203
bamboo, common, 187
bane, leopard's, 173; rat's, 189
baneberry, 55, 183; red, 55, 183; white, 55, 183
Barbados pride, 187
barbasco, 196
basswood, 205
bayonet, Spanish, 206
beadtree, 183
bean, castor, 104, 202; eastern coral, 176, 203; fava, 88, 206; hyacinth, 178, 196; java, 180, 200; kidney, 180; precatory, 77, 183; scarlet runner, 200; senna, 86; small black lima, 180; velvet, 179, 199; yam, 199
beech, 176, 193; American, 176, 193; European, 176, 193
beet, 187; sugar, 187
begonia, sand, 202, 47
belladonna, 148, 186
bindweed, 190
birch, 187
bird-of-paradise, 174, 187
birthwort, 186, 205
bittersweet, 159, 203; climbing, 189, 174
bitterweed, 194
black-eyed Susan, 202
blite, sea, 204
bloodberry, 202

bloodroot, 71, 203
bloodwort, 196
bluebells, California, 200
bluegrass, 200
boneset, 193
borage, 187
bouncing bet, 53, 203
bower, virgin's, 58, 190
boxwood, 107, 187; common, 187; Japanese, 187
bracken, 180, 201
Brazil beauty leaf, 188
bridal wreath, 204
broom, Scotch, 191
bryony, white, 174, 187
buckeye, 115, 184; bottlebrush, 184; painted, 184; red, 184; yellow, 184
buckthorn, 118, 202; Carolina, 202
buckwheat, 45, 193
bugbane, 190
bugloss, viper's, 192
bush, bellyache, 146; fetter, 184; jetberry, 181, 202; strawberry, 114, 193
buttercup, 62, 202; bulbous, 202; tall field, 202
butterweed, 203
buttonbush, 189

cactus, candelabra, 99, 193; rattail, 186
cajeput, 124, 198
caladium, 12, 187
calla, common, 207; wild, 187
camas, death, 32, 207; edible, 32
candelabra, spiny, 193
candlebush, 188
candlenut, 184
capers, 99
caraway, 188
carrot, cultivated, 192; wild, 131
cassava, 178, 198
cassia, ringworm, 188
catnip, 199
cattail, 205
cedar, Atlantic white, 189; eastern red, 7, 196; red, 7, 196; southern red, 196
celandine, 69, 189
celery, 173, 186
cestrum, 151
chard, Swiss, 187
chervil, 185; wild, 186
cherry, 75; black, 75, 201; choke, 201; cultivated, 201; ground, 200; Jerusalem, 159, 204; pin, 201; wild, 201
chestnut, horse, 115, 184
chickpea, 189
chicory, 189
chinaberry, 95, 198

Index of Common Names

cinnamon, 190
clover, 205; alslike, 205
clubmoss, running, 197
cockle, corn, 52, 184
cocklebur, 206; common, 206; spiny, 206
coffee, 190
cohosh, blue, 64, 189
columbine, European, 186
comfrey, common, 204
coontie, 5, 207
coriaria, 175, 190
corn, 207
cotton, 176, 194
cottonwood, 201
cowbane, spotted, 128
cowhage, 149
cowpea, 206
cowslip, 57, 188
coyotillo, 118
crabapple, southern wild, 198
crabgrass, 192
creeper, rangoon, 202; trumpet, 162, 188; Virginia, 120, 199
cress, garden, 197
crinum, southern swamp, 191
crocus, autumn, 23, 190; prairie, 201
crossvine, 185
croton, purging, 191; woolly, 98, 191
crowfoot, 62, 202; cursed, 202
crownflower, 188
crown-of-thorns, 193
cucumber, 191; creeping, 198; wild, 192
cudjoe-wood, 196
cycad, crozier, 3, 191
cypress, 191

daffodil, 36, 199
dandelion, 204
daphne, 175, 191
darnel, 178, 197
dasheen, 190
devil's bit, 189
dill, 185
dock, 47, 202; curled, 47
dodder, 191
dogbane, 186; spreading, 186
dogfennel, 193
dogwood, Jamaica, 200
donkey-eye, 199
dumbcane, 13, 192
Dutchman's breeches, 72, 192

ebony, green, 196
eggplant, 204
elder, box, 183; gulf, 203; marsh, 196; red-berried, 203
elderberry, 164, 202
elephant's ear, 190, 206; giant, 184; purple-stemmed, 206
elm, 205; English, 205
eucalyptus, 123, 193
evergreen, Chinese, 184

fennel, 194
fern, asparagus, 186; sweet, 190
fescue, 193
fetterbush, 197, 200
fig, 39, 194; creeping, 194
fir, balsam, 183
firethorn, 201
flag, 37
flax, 91, 197
flower, blanket, 194; cardinal, 166, 197; crepe, 196; leather, 58, 190; pasque, 185; slipper, 159; spider, 190; temple, 200; trumpet, 158, 203
flower fence, 187
foot, colt's, 205
foxglove, 161, 192
four o'clock, 49, 198
fritillary, snakeshead, 194

gallego, 201
garlic, 184
ginger, 207; wild, 186
ginkgo, 6
glasswort, 202
golden chain, 81, 196
golden club, 16, 199
golden drop, 175, 192
golden seal, 61, 195
goldenrod, 204
gourd, bitter, 165
grass, Bermuda, 191; orchard, 191; quack, 184; rye, 197; velvet, 195

hackberry, 189
hawthorn, 191
hazelnut, 191
hearts-a-bustin', 193
heath, climbing, 200
heliotrope, 177, 194
hellebore, 206; false, 30, 206
hemlock, ground, 8; poison, 130, 190; water, 128, 189
hemp, 42, 188; Indian, 186
henbane, 154, 195; black, 154
Hercules club, 125, 186
hickory, 188
hogfennel, 200
hogweed, 195; giant, 164, 177, 195
hogwort, 98
hollowstalk, 197
holly, 113, 195; English, 195; Florida, 110
honeysuckle, 197; Japanese, 163; trumpet, 197
hops, 43, 195
horseradish, 186
horsetail, 176, 192
horseweed, 190
huckleberry, garden, 204
hunter's robe, 12
hyacinth, 28, 195
hydrangea, 73, 195; garden, 73, 195; oak-leaf, 195; smooth, 73, 195

223

Index of Common Names

Indian pink, 139, 204
indigo, false, 174, 187; wild, 174, 187, 195
ipecac, wild, 193
iris, 37
ironweed, 199
ivy, Algerian, 194; English, 126, 194; ground, 194; poison, 111, 205

jack-in-the-pulpit, 11, 186
jessamine, 151, 189; Carolina, 137, 194; crape, 176, 193; day, 151, 189; night-blooming, 151, 189; willow-leaved, 189; yellow, 194
jimsonweed, 152, 155, 192
jonquil, 36, 199
juniper, 7

lambskill, 196
lantana, 146, 196
larch, European, 196
larkspur, 59, 192; rocket, 190
laurel, 196; cherry, 75, 201; Indian, 121, 188; mountain, 133, 196
laurel leaf, 190
leatherwood, 122, 192
leek, meadow, 184
lemon, 190
lettuce, water, 18, 200
lignum vitae, 177, 194
lilac, 204
lily, Amazon, 193; Barbados, 195; belladonna, 185; blackberry, 187; blood, 194; climbing, 194; glory, 26, 194; milk & wine, 35, 191; spider, 195; tiger, 197; zephyr, 207
lily-of-the-valley, 25, 190
lime, 92, 190
lobelia, great, 197
locust, black, 85, 202; clammy, 85, 202
loquat, 192
loveapple, 203
lupine, 82, 197; wild, 197

magnolia, southern, 198
malanga, 206
manchineel, 101, 195
mango, 108, 198
mangrove, black, 187
maple, 183
marigold, 204; marsh, 57, 188; wild, 204
marijuana, 42, 188
masterwort, 200
mastwood, 121
matasano, 188
mayapple, 65, 200
mayweed, 185
melilot, white, 198
milfoil, 167
milkbush, African, 204
milkweed, 143, 186, 196; butterfly, 143, 186; giant, 174, 188; scarlet, 186

mistletoe, 44, 200
monkshood, 54; garden, 54, 183
moonflower, 178, 195
moonseed, 67, 198
moonvine, 178, 195
morning glory, 145, 195; beach, 195; tall, 195
motherwort, 189, 197
mugwort, 186
mulberry, 199; paper, 187; red, 40, 199; white, 40, 199
mullein, 206
mustard, black, 187
myrtle, crepe, 196; wax, 199

narcissus, poet's, 199
nettle, Carolina horse, 203; horse, 203; spurge, 97, 190; stinging, 41, 206; wood, 196
nightshade, 159, 203; black, 159, 204; common, 159, 203; Brazilian, 204; deadly, 148, 187; graceful, 204; silver, 203
nosegay, 200
nut, bangor, 204; Brazil, 187; cashew, 151, 167; macadamia, 198; physic, 103, 196; purge, 206
nutmeg, 179, 199

oak, 201; poison, 111, 205
oats, 187
okra, 183
oleander, 142, 199; yellow, 181, 205
onion, wild, 184
oracle, beach, 187
orange, mock, 200; osage, 198; sour, 92, 190; trifoliate, 93, 201
orchid, lady's slipper, 38, 191

palm, bamboo, 189; false sago, 3, 191; fishtail, 10, 188; queen, 186
pansy, 206
papaya, 188
parsley, fool's, 130, 173, 184
parsnip, cow, 177; wild, 132, 199
pawpaw, 63, 186
pea, chick, 82; common, 200; everlasting, 83, 197; partridge, 188; rosary, 77; scurf, 201; sweet, 83, 197; wild winter, 197
peach, 75, 201
pear, 201; balsam, 165, 198; wild balsam, 165, 198
pennyroyal, 194; European, 179, 198
peony, 199
pepper, Brazilian, 110, 203; chili, 150, 188; marsh, 201
peppervine, 185
peregrina, 196
periwinkle, common, 206; greater, 206; Madagascar, 141, 189
philodendron, 17, 200; split leaf, 15, 198

Index of Common Names

pigweed, 185
pimpernel, scarlet, 185
pine, 200; Australian, 189
pineapple, 185
pinkroot, 139; West Indian, 204
pipsissiwa, 189
plant, air, 196; century, 33, 184; coral, 196; gas, 192; oyster, 20, 202; rice paper, 205; sensitive, 198; shoofly, 199
plantain, 200
plum, 75, 201; Ochrosia, 199
plumbago, 180, 200
pod, monkey, 200, 202; sickle, 189
poinciana, 174; false, 86
poinsettia, 99, 193
poisonwood, 109, 198
pokeberry, 50, 200
poke salet, 50, 200
pomegranate, 201
pongam, 180, 201
poppy, 70, 199; Mexican prickle, 68, 186; opium, 70, 199; rock, 69; white prickly, 186
pops, Christmas, 205
potato, 159, 204; air, 175, 192; devil's, 192
primrose, 135, 201
privet, 136, 197
purple queen, 21, 205
purse, shepherd's, 188
purslane, 180, 201; sea, 205

quarters, lamb's, 189
Queen Anne's lace, 131, 192

radish, cultivated, 180, 202; wild, 181, 202
ragweed, 185; lance-leaved, 185
ragwort, tansy, 172, 204
raintree, 202
rattlebox, 79, 191
rattlebush, 203
rattleroot, 190
redbird-cactus, 179, 199
redbud, eastern, 189
rhododendron, 134, 202
rhubarb, 46, 202
robe, hunter's, 14, 192
rocket, London, 203
root, Culver's, 206; dragon, 186; papoose, 64; yaw, 204
rice, 199
rose, 202; cabbage, 202; Christmas, 60, 195; damask, 202; French, 202; tea, 202
rosemary, 202
rubberplant, 194
rubbervine, Madagascar, 144, 191; Palay, 144, 191
rue, 94, 202; goat's, 87, 205; meadow, 205
rush, 196
rye, 203

saffron, meadow, 23
sage, 202
sage, white, 186
salad, rocket, 192
saltbush, 187
saltgrass, 192
sandalwood, red, 183
sapodilla, 179, 198
sapote, white, 188
seal, Solomon's, 201
sedge, 188; sea, 188
senecio, 170
senna, 188; Christmas, 188; coffee, 78, 189; wild, 188
sesbane, purple, 86, 203
seven bark, 195
shower, golden, 188
silkoak, Australian, 194
skunk cabbage, 19, 204
slipper, pink lady's, 191; yellow lady's, 191; small white lady's, 191; showy lady's, 191
smartweed, 201; dotted, 201
snailseed, 190
snakeroot, black, 32; Seneca, 201; Virginia, 186; white, 168, 184
snapdragon, 186
sneezeweed, 169, 194
snowberry, 204
snow-on-the-mountain, 99, 193
soapberry, 203
sorrel, 47, 202; garden, 202; wood, 180, 199
soursop, 185
spearmint, 198; Scotch, 198
spinach, 204; New Zealand, 205
spurge, 99, 189, 193; caper, 99, 193; cypress, 193; flowering, 193; leafy, 193; red, 193; spotted, 193
staggerbush, 197
star-of-Bethlehem, 29, 199
stonecrop, mossy, 203
sumac, poison, 111, 205
sundew, 192; roundleaved, 192
supplejack, 187
sweetgum, 197
sycamore, 200

tailflower, 186
tansy, 172, 204
tares, 178
tea, Laborador, 197; marsh, 197; Mexican, 48, 189
terminalia, Mueller's, 205
thoroughwort, 198
thyme, garden, 182, 205
timothy, 200
toadflax, 197
tobacco, 156, 199; Indian, 166, 197; tree, 199; wild, 156
tomato, 155, 197
tongue, adder's, 193; hound's, 191

Index of Common Names

top, red, 184
tree, bean, 189; ben, 199; be-still, 181; bottlebrush, 124; cabbage bark, 185; calabash, 191; camphor, 190; cashew, 173, 185; catawba, 189; coral, 176, 193; corkwood, 203; fringe, 189; hemlock, 130; Japan wood-oil, 184; kassod, 189; Kentucky coffee, 80, 194; mu-oil, 184; muscle, 188; Panama, 204; pencil, 99, 193; popcorn, 106, 203; sandbox, 102, 195; spindle, 114, 193; tung oil, 96, 184
tree-of-heaven, 184
trumpet, angel's, 187
tulip, 205
turnip, 187

vanilla, Mexican, 206
varebell, 198
vetch, 89, 206
vine, balloon, 188; chalice, 158; cow-itch, 162, 188; garlic, 201; love, 189; matrimony, 197; Mexican flame, 170, 203; wait-a-bit, 187
violet, English, 206

wahoo, eastern, 114
walnut, black, 146

watercress, 199
weed, bishop's, 185; guinea hen, 200; Jamestown, 152; maddog, 184; puncture, 205; scorpion, 200; squaw, 170, 203
wicky, 196
wicopy, 122, 192
willow, 202
windflower, 56, 185
wintergreen, 194
wisteria, 90, 206; American, 206; Chinese, 206; Japanese, 206; Kentucky, 206
wolfbane, 54
wormseed, 48, 186
wormwood, 48; Louisiana, 186
wort, St. John's, 195
wreath, bridal, 204

yam, giant coco, 206
yarrow, 167, 183
yaupon, 113, 195
yew, 8, 205; American, 205; English, 8, 205; Florida, 8, 205; Japanese, 8, 9, 200, 205